HALLEY'S COMET, 1910:
Fire in the Sky

Recreated by Jerred Metz

Singing Bone Press
St. Louis, Missouri

Hardback: ISBN 0-933439-00-8
Softbound: ISBN 0-933439-01-6

Book Design and Cover: Tom Lang
Composition: Top Graphics;
Production, Joanne Kluba;
Typesetting, Terry Prutzman
Printing:
Christian Board of Publication

Halley's Comet, 1910: Fire in the Sky is available at special quantity discounts for bulk purchases for sales promotions, premiums, fund raising or educational use. Special books or book excerpts can also be created to fit specific needs.

For details write Singing Bone Press, Department B, P.O. Box 1650, St. Louis, MO 63188.

DEDICATION
R. BUCKMINSTER FULLER
(July 12, 1895, Milton, Massachusetts - July 1, 1983)

I was fifteen at the time of Halley's comet's appearance. I lived in Milton, Massachusetts and attended Milton Academy. It was an extraordinary school. At that time it was exclusively a Harvard preparatory school. It was at the highest rank scholastically of all the Harvard preparatory schools. So things like Halley's comet were of great interest and were discussed in the very best scientific way. We were all excited.

My father had a beautiful camera for those days, a Kodak Bull's Eye. It took 4" x 5" film. There were not really many good cameras like that in those days.

The comet was visible for quite a few evenings. I'm sure I didn't take the photograph the first evening. I remember coming out from the house to take the picture quite early in the evening. We had quite a wooded place at our house. The comet was showing between a big pine above some oaks. I was able to get a good view of it. I got a very good picture.

Nobody in my community was superstitious about the comet. We knew it was coming and there it was. It was just a normal phenomenon, like everything else.

** * **

When I was twenty-eight Hubble discovered another galaxy other than our Milky Way. Up to six months ago we had discovered two billion more galaxies and within the last six months we had discovered two hundred billion more galaxies with the radio telescope.*

I have seen remarkable changes in my lifetime.

**Dedicated to R. Buckminster Fuller,
Werner Erhard,
and
The est Training**

*Conversation held on April 3, 1983.

TABLE OF CONTENTS

ACKNOWLEDGEMENTS

It is a great pleasure to acknowledge the contribution of many people whose knowledge and assistance, so willingly shared, made this book possible.

First and foremost, is Ruth Freitag, Research Librarian, Science and Technology Section, Library of Congress, whose bibliography of Halley's comet material housed in the Library made much of this book possible. Her vast knowlege of the Library of Congress' holdings and her masterful skills as a researcher are gratefully acknowledged.

Joe Laufer, founder, publisher and editor of *Halley's Comet Watch '86* has been a friend of this project since its early days, as has Fred Shaaf, founder and publisher of *Dark Skies for Comet Halley* (DSCH).

Other organizations have supported specific phases of the book's writing and production:

The Virginia Center for the Creative Arts provided a setting where much of the book was written; The St. Louis Public Library staff; the Missouri Historical Society; Astronomical League of America; *Astronomy Magazine* and the Library of Congress, Photoduplication Department.

The book's legion of supporters whose encouragement is so greatly valued include Tom Lang and Phil Sultz of Singing Bone Press, Bob Vogel, Howard Schwartz, Anne Keefe of KMOX-AM, Lorin Cuoco of KWMU-FM, Barb Selders, Joanne Kluba, Walter and Sarah Jones, and Wayne and Sharon Salomon. I am especially thankful to Joyce Flaherty, literary agent, whose enthusiasm for the book kindled my own. I thank my family and friends who lent financial support to the Halley's comet project.

And I thank my partners in life, Sarah and Ravenna Barker and Zachary Metz, for giving me the time and encouragement to write.

INTRODUCTION

Halley's Comet, 1910: Fire in the Sky is not an astronomy book in the traditional sense, but a book in which an astronomical event is the central character. Halley's comet comes once in a long life-span, about 76 years. The last time the comet was visible to the naked eye and field glass was in April, May, and June, 1910. And on the night of May 18, when the earth was to pass through the comet's tail, *the greatest number of people ever to share in a single event watched and waited with mingled fear and awe.* The book documents the spectacle of humanity's reaction to the comet and recreates the stir it caused. The able writing of by-line and staff writers, correspondents and stringers, the music of composers and songwriters, illustrations from sheet music cover artists and post card artists, ads from companies selling soap to champagne—all appearing in a multitude of newspapers, magazines, and journals—made this book possible. The freshness and liveliness of the stories arises from the genuine excitement the writers and artists felt and captured, some of which is recreated here. So these remarks are to pay tribute to the writers, artists, thinkers, scientists, and commentators of seventy-five years ago who turned their thoughts and gaze to Halley's comet. I credit the sources in the text and in section to be found at the end of the book.

The book includes news reports, human interest stories, "yellow" journalism, scare pamphlets, advertisements, letters, editorials, humor, broadsides, poems, cartoons, and even sports stories about the comet. There is little here of the reporting of history or astronomy in the ordinary sense. Instead there are *stories* of people and their activities in response to the comet's presence. The book presents a "slide show" of humanity seen through the "magic lantern" of the comet's 1910 apparition.

Welcome to the show.

I
Astronomy

COMET HALLEY, 1910
Used with permission of Lowell Observatory

1. WHO WAS HALLEY (b. Nov. 8, 1656, d. Jan. 14, 1742) AND WHY IS A COMET NAMED FOR HIM?

Edmond Halley's discovery of the comet named for him is a landmark in the history of astronomy. Halley, one of the most versatile minds of his day, is chiefly remembered today for discovering a comet's *return* along a path which he was the first to calculate. The story of Halley's remarkable discovery and his other contributions to science will be recounted again and again in the popular press during the next several years. Until Halley's calculation, it was believed that comets appeared once, never to return again.

Sir Isaac Newton's *Principia Mathematica,* the great work on gravitation, "the mistress of all motion in the universe," had laid in his desk for a long time though it was complete. Halley, a personal friend of Newton, asked to read it. Realizing the immense significance of the work, Halley asked his friend to allow him to publish it. In 1686 the *Principia Mathematica* was published at Halley's expense and introduced by a poem he wrote in Latin. Newton demonstrated that every body in the universe attracts every other body directly proportional to their masses and inversely proportional to the square of their distances. All motions in the heavens are governed by this law.

Reasoning that comets must move in accordance with Newton's principle, Halley computed the courses of twenty-four recorded in chronicles and discovered that they all described parabolas around the sun, as Newton's law necessitated. Among the twenty-four were three which took approximately the same course. These were the 1682 comet, which Halley himself had seen, and comets of 1607 and 1531. The intervals between the three appearances were nearly the same. Halley deduced that because these comets followed paths at the same intervals and traveled in an *ellipse* instead of a parabola, the three appearances were of the same comet. He boldly predicted that in the end of 1758 or in the beginning of 1759 the comet could again be visible from earth. Halley's discovery of the behavior of this comet is a landmark in the history of astronomy.

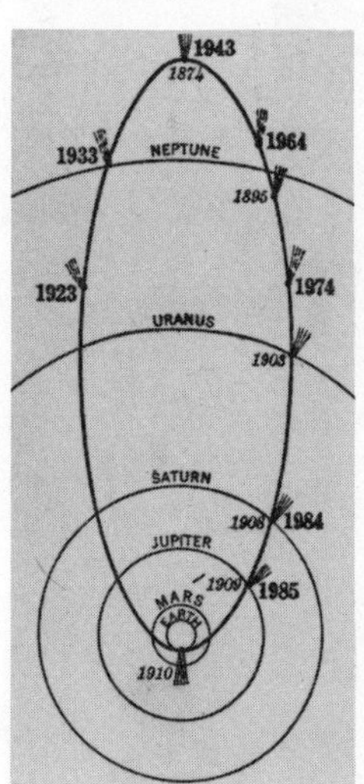

This simple sky map locates Halley's comet through two entire orbits, 1835 through 1910 and 1910 through 1986. The 1986 apparition will be the twenty-ninth appearance of the comet to be noted by astronomers. It will be the fourth since Edmond Halley set the milestone of celestial mechanics by determining that this comet had been seen before and that its return was certain.

2. HALLEY'S COMET AND ITS GREAT JOURNEY

An introduction to Halley's comet appeared in *St. Nicholas Magazine* for children in which the prominent features and physical laws the comet would display were described in such a way that fifth- through twelfth-graders could understand them.

We know all about the comet now, just when it will come again, and where it goes between visits. We know that it travels around the sun just as we do, but in an oval or elliptical path, with one end of the ellipse only half as far and the other thirty-five times as far from the sun as we are....

Usually a comet has a bright spot in the head which is called the "nucleus." This is made up of solid matter. Around this is a hazy mass called the "coma" or "hair," which is made up of gas. This gas streams at one side into a misty, gauze-like tail. But comets look different at different times. Could you have seen Halley's comet thirty-seven years ago, when it was at the other end of its ellipse, 3,300,000,000 miles off, you would not have recognized the small, dark object, colder than the coldest ice, without the slightest trace of a tail, with no light of its own, traveling in a region of perpetual twilight, where the sun looks no brighter than a street lamp. Since then the comet has been moving toward the sun, very slowly at first, then faster and faster, until by the twentieth of April it was traveling at a speed that would have made a rifle-shot look as if it were not moving. When it was first picked up by the largest telescopes last fall, it looked like a mere speck; but as it approached the sun, the cold, frozen comet material began to warm up. Then it began to glow, and, like coal when heated, it commenced to give off volumes of gas that streamed out into a tail....

On the twentieth of April the comet rounded the sun and began its return journey, back to the cold, dark regions it came from. As the sun was passed, the light, flimsy tail swept around faster than the head, and the comet is now moving off tail first.

During April and the first part of May the comet could be seen only before sunrise. After the eighteenth of May it appeared in the evening skies.... People back in the shadow just off this thread of light can peer over the eastern horizon and see the comet without getting the glare of the sun, but as the earth turns on its axis it carries them around into the glare of the sun, and even before they get into the direct rays the comet will fade from view in the twilight of approaching dawn. In the meantime the earth is traveling bodily toward the right, as shown by the dotted line, while the comet is moving toward us. After the eighteenth, when the comet passes between the earth and the sun, it crosses over to the other side of the earth, and there it could be seen after sunset by peering around the *western* horizon at it ....Each day, as it moved away the comet could be seen later in the evening, growing fainter and smaller until, almost at the end of June, it was no longer visible to the naked eye. The big telescopes followed it much farther on its journey. In time they lost sight of it. The comet pursued its lonesome journey to the cold dark regions whence it came.

From "Halley's Comet" by A. Russell Bond
in St. Nicholas Magazine

After the 18th of May the comet appeared in the evening sky. Illustration from St. Nicholas Magazine.

The Comet's Speed

The comet's speed is affected by the immense gravitational forces exerted by the planets it passes on its journey, especially Jupiter and Saturn. For this reason the comet's orbit is not precisely seventy-five years, as most believe, but has been as long as seventy-nine and as short as seventy-four years. Astronomers call these disturbances "perturbations."

Due to this force of gravitation, in the *single year* from October 1909, to October 1910, Halley's comet traveled as great a distance as it had during the *twenty-five years* from 1860 to 1885. In the *month* of April 1910, when nearest the sun, Halley's comet traveled as far as it had during the *four years* from 1870 to 1874 when the comet was farthest from the sun.

If the comet was traveling west and crossed the Hudson River just as a bullet was fired from a rifle, by the time the bullet reached the other side the comet would have passed Philadelphia. In other words, the comet traveled at a speed of 43 miles per second!

An editorial in the *Humanitarian Review* observed that the comet and the earth would pass each other at a combined speed "three thousand times faster than our most rapid express trains move." The comet "will be moving in its orbit at 43 miles per second—its most rapid movement." From there its velocity slowly decreased until it reached apehelion, the farthest point from the sun. It was moving so slowly that it began to fall towards the sun. And, again, its velocity will increase on its return trip achieving maximum speed when it arrives at perihelion in 1986.

Major Events in the 1910 Apparition

1. August 29, 1909—comet recovered by Max Wolf, celestial photographer, at Köningstuh Observatory, Heidelberg, Germany.
2. February, 1910—tail of 1° detected comet seen in conjunction with Venus from earth. This is an extremely rare occurrence especially since the comet does not travel in the same elliptic as the planets in the solar system.
3. April 20, 1910—the comet rounds the sun and begins its return journey to the outer reaches of the solar system.
4. Until May 18, 1910 the comet is visible only in the east just before sunrise.
5. May 18, 1910—the comet transits the sun and crosses a line between the earth and the sun and across the sun's face.
6. May 18, 1910—the earth was to pass through the comet's tail.
7. After May 18, the comet was seen only after sunset in the west.

* * *

The comet's appearance has coincided with events central to civilization's history. In 1910, writers often presented these coincidences as being the *result* of the comet's presence. Tracing the comet's path through history, we find it visible at many remarkable events. In 65-66 A.D. the comet was seen by St. Peter and the Jewish historian, Josephus, over the city of Jerusalem just before the fall of the Holy City to the Romans and the destruction of the second temple. It was present during the Battle of Hastings in 1066 when William the Conqueror and his Norman army crossed the English Channel defeating King Harold and his Saxon army, thus changing the course of British history. During the year 1222, the comet was visible during September and October. In those months and immediately afterwards, Genghis Khan, the bloody Mongol conqueror, and his fierce hordes ravaged all of China, Persia, India, and the Caucasus as far as the River Don. Khan believed the comet was a portent signaling for his forthcoming conquest of the world.

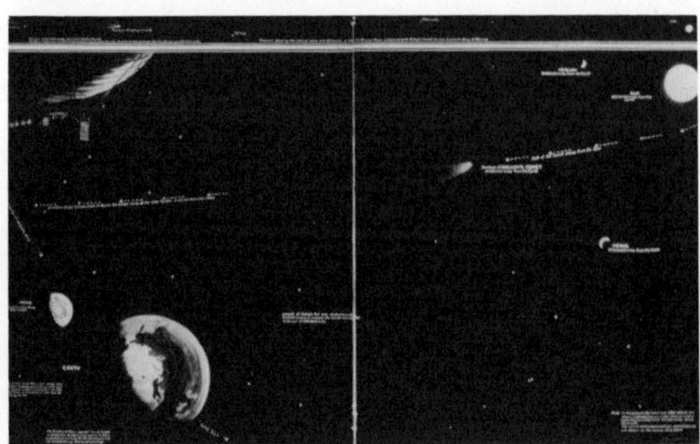

COLLIDING WITH A COMET'S TAIL: A celestial spectacle which may be enacted next week between our earth and Halley's comet.

The London Graphic, *a magazine catering to the English nobility, portrayed the comet and other details shown as if seen from an dirigible 200,000 miles above the earth. The perspective added a wonderful touch of science fiction to the sky map, as if to say that by 1986 there could be space travel.*

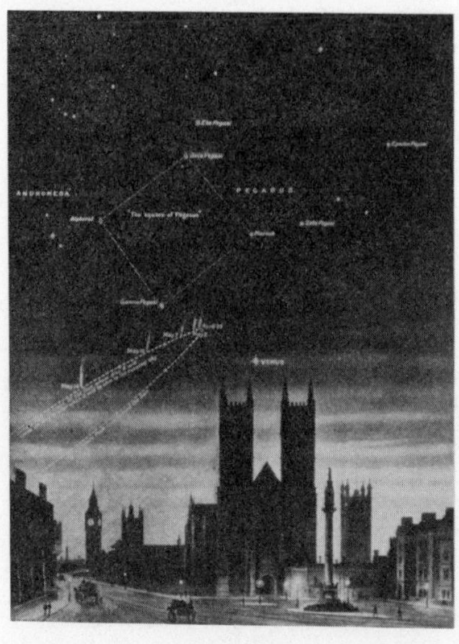

Where to look for Halley's comet in the morning sky in 1910.

Dozens of sky maps, ephemerides, or schedules showing the comet's computed daily place in the heavens were printed.

3. THE COMET'S POISONOUS TAIL:
THE "TERROR OCCASIONED BY THE NEAR APPROACH OF HALLEY'S COMET"

To appreciate the "terror occasioned by the near approach of Halley's comet" only a few facts are needed. Some people were afraid the comet would collide with the earth. Superstitions and beliefs led others to see the comet as an omen of coming disaster. But most people who were afraid of Halley's comet in 1910 were in fear of its *poison tail.*

In the enlightened world of 1910 it was believed that astronomy's certainty would bring superstitions about the heavens to an end. But about Halley's comet, astronomy was certain that the tail contained cyanogen. When mixed with hydrogen, cyanogen produces the dreadful poison, prussic acid. Only a grain of this potassium salt *touched* to the tongue brings instant death! In its uncombined state, cyanogen is a bluish gas similar in chemical behavior to chlorine. Its odor resembles that of almonds.

The tail contained a minute quantity of the gas. The scientists—astronomers, biologists, chemists, and toxicologists—assured the world that there was no possibility of any harm coming to anyone or anything. The facts were spread as far and wide as newspapers, magazines, pamphlets, and word of mouth could reach. Most people grasped the idea, but many experienced intense anxiety and dread. The reported presence of cyanogen was enough to create terror in the minds of the ill-informed.

On February 6, 1910 the director of Yerkes Observatory and his assistants trained the telescope on the comet to examine the spectrum of its gases. The light coming through the telescope was refracted through a set of glass prisms, the comet's light being projected onto a screen. By comparing the vari-colored lines of the comet's spectrum with the lines given by specific gases, the scien-

6

From Collier's Magazine

tists could identify cyanogen, a gas commonly found in comets' tails. The Daylight Comet, also known as Comet 1910,A, visible in January, had contained it, and cyanogen was detected in many others as well. The difference was that the earth was predicted to pass through the tail of Halley's comet on May 18. The discovery of the cyanogen on February 6 was communicated to other astronomers as well as to the public. The scientific view was that the cyanogen would have *no effect whatsoever* upon life on earth. Professor W. J. Hussey of the University of Michigan discounted the idea of any danger to life resulting when the earth passed through the tail:

> All comets are of the same chemical nature. The earth has passed through comets' tails before and no bad effects have been felt. The comet is 14,000,000 miles away, and the cyanogen gas from it will not be sufficient to produce any noticable effect. It will be like adding one molecule of gas to a large roomful of air. There cannot be any possible danger even to an insect from the small quantity of hydrocyanic acid which might result in the atmosphere. You will find fifty times more of the same gas in any chemical laboratory.

On February 11 a *New York Times* editorial entitled "Poison in the Tail of a Comet" seconded Professor Hussey's assurances:

People who were disturbed by the recent news that the "cyanogen line" had been formed in the spectrum of the tail of Halley's comet by the astronomers at the Yerkes Observatory have now been kindly relieved of their anxiety by Prof. Hussey of the University of Michigan. He does not deny that cyanogen is a most terrible poison or that a comet would not have to add much of it to our atmosphere to destroy very promptly all terrestrial life. He only calls attention to the fact—which anybody, by the way, can find in any recent astronomical work—that the tails of comets are of such almost inconceivable tenuity, and that even though they were composed of nothing but cyanogen the earth could pass through a dozen of them without producing the slightest effect upon the most delicate of its inhabitants.

Cometary tails are so near to nothing at all that the thinnest mist of which we have any knowledge is grossly material in comparison. The spectroscope's verdict is always beyond question and wherever it finds cyanogen or anything else there the thing indicated by the "lines" undoubtedly exists, but the discovery, though sure, may mean amazingly little as to quantity, and such is undoubtedly the case in regard to comets. Only to ignorance or superstition are they alarming.

Explaining it another way, Professor Mitchell of Columbia University claimed that the number of particles per cubic mile in a comet's tail is so small, if a cubic mile of it could be concentrated in a glass beaker in the laboratory the greatest refinement of chemical research would barely be able to detect any cyanogen at all. He also noted that the earth's atmosphere is thousands of times denser than the comet's tail. The particles of cyanogen could not penetrate the atmosphere and reach the earth's surface.

In an address in Boston on March 21 Professor Percival Lowell, astronomer and director of the Flagstaff Observatory, Flagstaff, Arizona, described the tail of Halley's comet as "The airiest approach to nothing set in the midst of naught…"

"On account of its vacuity it could have no injurious effect on humanity, even if it were the deadliest of gases," said Prof. Lowell, in disputing the theory that danger to the earth's inhabitants from the comet is possible. The nature of astronomy was such that there was the possibility of great accuracy in observation and prediction *and* erroneous speculation at the same time. Even Professor Lowell, correct on the matter of the comet's tail, sounded less than scientific on some matters. He left Boston to lecture successfully in London and Paris on the Martian canals and propounded his theory that they were built by intelligent beings. Some astronomers did announce that contact between the earth and the comet's tail would bring animal life on earth to an end. A French astronomer, Professor Deslandres of the University of Dijon, said, "The hypothesis that the gas was liable to affect the terrestrial atmosphere would not be at all absurd." Noted French astronomer M. Camille Flammarion predicted that, *The cyanogen gas would impregnate the atmosphere and possibly snuff out all life on the planet.* In addition to writing accurate, clear reports on the comet's activities during April and May of 1910 for the popular French news magazine, *L'Illustration,* Flammarion convinced some that the comet posed a grave danger.

May 18, 1910 was a day, of dread. The few incidents of hysteria, madness and suicide occurred on or about May 18 because of the terror aroused by the passage through the comet's tail.

Datelined May 17, this article describes the fear at its greatest pitch in one city:

CHICAGO IS TERRIFIED

Women Are Stopping Up Doors and
Windows to Keep Out Cyanogen

Special to *The New York Times.*

> Chicago, May 17 - Terror occasioned by the near approach of Halley's comet has seized hold of a large part of the population of Chicago. Especially has the feminine portion succumbed. All else is forgotten. Comets and their ways and habits have been the principal topic discussed in the streets, cars, and elevated trains today.
>
> Telephones in the newspaper offices have been busy all day with inquiries as to the latest developments in the foreign visitor in the heavens. The principal fear is not that the comet will strike the earth, but that the gas which it is supposed makes up the tail will wipe out all life.
>
> "I have stopped all the windows and doors in my flat to keep the gas out," said one woman over the telephone. "All the other women in the building think it is a good thing, and all are doing the same. Do you think it will keep out the gas?"
>
> Physicians say that there were scores of calls today for their services from women who were suffering from hysteria following the reading of the possibilities of danger from collision or from the dreaded gas.
>
> "My patients say that they are nervous and are suffering from a feeling of depression," said one doctor this afternoon. "They add that they have been reading about the comet and I know that they are simply frightened."
>
> Others calling over the telephone have asked if it is probable "That we will all die at the same time, if the gas is strong enough!" and, "What is the exact time that the first of the gas will strike Chicago, or will New York get it first!"

How relieved those many people who trembled in fear throughout the days before May 18th must have been when they awoke on the 19th to find the world no different than before. Thanksgiving replaced fear and, as the astronomers knew it would, life went on as always. Except, of course, the comet had wrought some differences, and those are the stories this book tells.

4. 1910: THE YEAR OF CELESTIAL FIREWORKS

As remarkable an event as the appearance of Halley's comet is, the 1910 apparition was accompanied by other events visible in the heavens which, in the slow working of astronomical phenomena, could be considered the equivalent of celestial fireworks. On January 16 and on March 3, two new comets were discovered, each spectacular and unusual in its own way. On May 18 a monumental sequence of sunspots erupted on the sun's surface outside the normal eleven and one-third year pattern of recurrence. Accompanying the comet's own appearance were anticipated events including meteorite showers and auroras. The grand finale of this succession of unusual scenes in the heavens was a total lunar eclipse, which occurred twenty-two minutes after the comet set on the night of May 23.

One of the earliest Delft commemorative plates portrayed Halley's comet and Comet 1910, A. From the collection of Violette Steinke.

Comet A, 1910

On January 16, 1910, an astronomer in South Africa discovered a new comet conspicuously bright and visible to the naked eye. Harvard College Observatory received a cablegram: "Comet A was observed by Innes at Johannesburg on Jan. 16, .8110 G.M.T..." There are only four or five comets in a century bright enough to be seen during the day; this was one. At Lick Observatory the Daylight Comet was seen with the naked eye at noon, and it was a brilliant object in the southwesterly sky immediately after sunset.

By January 19 the comet appeared in the position of the constellation Capricornus. It was easily seen by the naked eye, shining brighter than Venus and with an impressive tail. Observatories in southern France and Algeria said the comet appeared with dazzling brilliance for several minutes on the night of January 20. The comet was seen by observatories in Wilhelmshave, Prussia, and Berlin, Germany. Comet A, 1910 outshone Halley's comet in popular interest. Many mistook it for Halley's.

The New York Times editorialized with ponderous humor:

CONFUSING THE COMETS

The comet some clear-sighted folks say they have seen near the western horizon of late is not Halley's comet, but an unidentified butter-in from space, now called for convenience, "A, 1910," which is as good a name as any for a comet without a history. It may be older than Halley's comet, but is not closely identified with the delusions of man, or his slow development in science. The only comet in the heavens now visible to the naked eye is the intruder called "A, 1910." It may have been "A, 1910 B.C.," for all we know, but it is new to us, and as a stranger we give it welcome. But we are justified in regarding it with some suspicion. It comes unheralded, and it demands a great deal of careful investigation. We may well doubt its possession of evil powers, but we must not confuse it with a steady-going old comet like Halley's, which has long outlived the misdeeds of its youth, for which it was not entirely responsible, and comes into view every seventy-five years or so with the same cheerful aspect. Until a new Federal bureau is established at Washington to see that all comets are properly branded with the owner's name, we must be cautious.

Each evening millions of eyes eagerly gazed at the comet. No comet of this magnitude had been seen since 1872. The brightness of the nucleus equaled that of Mercury, and the tail measured four times the moon's breadth.

The comet was first visible along the east coast of the United States on January 21. On roofs of apartment houses and homes in the northern part of New York City, at windows in hotels and in the offices of downtown buildings thousands gathered in the late afternoon to wait and watch. A low flying bank of clouds hovered over the New Jersey skyline as the sun neared the horizon. The sun disappeared, the cloud bank seeming to follow. To the naked eye the comet's tail appeared at first to be a long, broad, single band, a shaft of light dull in color pointing straight upward into the sky. But, as approaching darkness brought the comet more prominently into view, the shaft was revealed to be parallel streams of light. The head came into view, flickering an instant just above the rim of the clouds, then disappeared.

The splendor of the spectacle justified the assertion of previous observers in other parts of the world; Comet A, 1910 was probably the finest comet seen within the past hundred years.

On January 23 Harvard Observatory released this statement:

We are now thoroughly convinced that this is an absolutely new comet, as we cannot identify it with any other known to astronomical records.

Harvard classified the comet as A, 1910, the letter signifying the order of its discovery during the calendar year which makes up the rest of its name. It was also referred to as Drake's or Innes' comet.

By January 24 the comet was already receding from earth's vicinity, traveling at more than 120 miles a second. A, 1910 lit part of the voyage of the steamer, *J. R. Stetson,* from San Pedro, California. When he arrived at San Francisco on the 25th, Captain Bonifield said the comet was of great brilliancy for an hour.

The next day the tail split in two. Electrical disturbances occurring in the comet just before it broke apart caused the tail to split. Spectrographic analysis of the comet's composition revealed that it was the presence of sodium which accounted for the comet's great brilliance. Sodium is very rare in comets which are usually composed of hydrocarbons. The *Times* took this discovery as an occasion to let some humorist's wit take to the field in an editorial.

BOYCOTT THE SALT COMET

The comet labeled A, 1910, which is not included in the authentic catalogue of 22 identified comets, 14 with periods ranging from 4 to 14 years, and 8 with periods of from 32 to 76 years, is causing a great deal of perplexity, as might have been expected. Few have seen it, but all know it is there, as Mr. George Sampson remarked of Miss Lavinia Wilfer's flannel petticoat. Those who have seen it pronounce strange things concerning it. One spectroscopic artist has discovered sodium, an element hitherto lacking in well-ordered comets, in A, 1910. It is just possible that some alert scientist has been putting salt on its tail, but it is more likely that this strangely perverse comet has a lot of things in its composition which do not belong to other comets. A salt comet is certainly a novelty. No wonder the simple, credulous Mexicans are scared. It has two tails, one of them of extraordinary length! Its behavior from the beginning has been unlike that of the catalogued comets.

We do not believe for a moment that its appearance should cause any apprehension, because comets are all arrant humbugs, the tails of which do not blaze at all, but only seem to, while their heads are as light as known to the best society, and their bodies non-existent. But it would serve this comet right if it were wholly ignored. That would teach it a lesson. This year of 1910 belongs exclusively, according to the astronomers, to the big comet called Halley's. The programme of performances this summer by Halley's comet was arranged more than seventy years ago, and appeared in the textbooks when one's grandfather went to school. It will be carried out with no postponement on account of indisposition or the weather.

But some months before the scheduled time for Halley's comet to reveal itself to the unaided eye this double-tailed, well-salted hobo of the heavens wanders into our ken and rashly assumes responsibility for the floods and fires and strikes and the high cost of living, without any historical authority whatever. The wine of 1910 will be remarkably good, because comet-year wine is always better than that of other years, and Halley's comet was prepared to take the credit for it. We know what Halley's comet has already done for the world, since it outgrew its old evil ways. Let us stick to Halley's and boycott the unheralded comet!

Comet B, 1910, or Pidoux's Comet

Comet B, 1910, the second discovered that year, was truly an astronomical oddity. When discovered on March 3 by Geneva Observatory, its position was right ascension, 40 minutes, 22 seconds; declination, 4 degrees, 51 minutes north, an observational position *nearly identical with Halley's comet.* Appearing in the same visual position as Halley's, the comet was thought at Greenwich Observatory to be an error, but two days later the director, the eminent astronomer A. C. D. Crommelin, asserted that it was genuine and a different comet altogether.

One commentator noted, "Its (Pidoux's comet) fleeting observations were followed by Socialist franchise riots in Germany and by Labor riots in Philadelphia, with widespread bloodshed between the rioters and the constabulary."

Halley's Comet and Related Phenomena

On May 17 a series of events began, some caused by the comet and others believed to be. In Tacoma, Washington, comet-watchers observed an interesting phenomenon. Between 2:15 and 3:00 in the morning the Milky Way, at that time quite close to the horizon, had the appearance of a bank of fleecy clouds through which the stars shone with twice their usual brilliance. The earth seemed enveloped in an odd sort of haze at sunrise, the first sign of which was one white ray that shot across the city to the southeast of Mt. Rainier.

That same night Professor D. W. Morehouse, head of the department of astronomy at Drake University in Des Moines, Iowa, reported a brilliant meteorite shower at 2:30 a.m. in the east.

The next day there were widespread activities in the sky outrivaling all expectations. Of course there were the anticipated auroras, meteorites, and lights in the sky. Part of a meteor was found in the office of J. W. Going, a real estate agent, at Jackson and Seventh Streets, Topeka, Kansas. It was hot when it fell. Washburn College authorities said that it had not come from Halley's comet. And there were events which no one had anticipated or could readily explain. At Yerkes Observatory in Wisconsin a spectacular display of aurora lights reached across the sky from east to west. Professor Frost did not connect the phenomenon directly with the comet. That night in Seattle a sight confronted Professor J. E. Gould of the university observatory. The professor was on his way home across the campus about 9:30 when he glanced at the eastern sky and stood in his tracks. "I could scarcely believe my eyes," he said afterwards. "There was the broad tail of Halley's comet fully 25 degrees long and four wide. The head wasn't due to rise until about three in the morning. It was a magnificent sight. It should be something tremendous later in the night when more of it is above the horizon." The sight indicated the splendor people could expect to see in the evening sky after the 18th. Observers in Lille, France saw pink and violet lights the same evening.

Sun Spots

The biggest surprise occurred on the sun. Headlines announced:

SUN SPOTS APPEAR; NOT DUE TO COMET?

One, Probably 150,000 Miles in
Diameter, St. Louis
Astronomer Reports

NEED OCCASION NO ALARM

GIGANTIC SPOTS
BLOT DAY STAR

Seattle Post-Intelligencer

Of course astronomers were certain the sunspots were caused by unusual internal activity, but the newspapers needed to hint at a connection between Halley's comet and the sun spots. Here is what happened. On the afternoon of May 18th, Rev. Martin Brennan, Professor of Astronomy at Kenrick Seminary, St. Louis, discovered extensive sun spots.

> There are three awful groups. In one of the groups I counted twenty-six spots. There is a large isolated one near the bottom of the sun. In the principal group one of the spots is the largest I have ever seen. It is 150,000 miles across. The recurrence of spots at this time is extraordinary.

The sun spots were unexpected because the usual period of recurrence is eleven and one-third years, and the most recent occurrence had concluded three years earlier.

Rev. Brennan also said,

> The comet is 80,000,000 miles away from the sun. There is no possible connection between the comet and the sun spots. It would require a body as large as the earth falling into the sun to make a spot as large as the major one in the principal group.
>
> The biggest sun spot ever recorded was seen by Capt. Dabis in August, 1848. It was 183,000 miles in diameter. The one I saw today I witnessed just before and while it was breaking. Near the bottom appeared a large rent as if the photosphere was torn.

In the evening of May 18th Prof. Jerome Recard of Santa Clara College Observatory issued the following statement:

> After nearly two months of rest, the solar surface is showing a recrudesence of activity well worthy of a maximum period. On May 18, at 1:00 p.m., there could be seen a large, intensely bluecolored spot, convex to the westward, concave to the eastward, in shape nearly like a half-moon.
>
> As a master spot it had a retinue of fourteen little ones, following in the rear on the eastern side, led by a vanguard of one taller and bigger than the rest. At a distance of a few degrees stood another group of three spots, a big one and two small ones.

The largest black spot measured 150,313 by 78,773 miles. The smaller group appeared on May 12 and was single then became triple. The larger group was first seen on May 15 on the eastern limb as a family of seven, then of nine on May 16, of eight on the 17th, and finally of fifteen on the 18th.

Why this sudden change on the star of day? It may be held, it has been well nigh demonstrated, that the rise and the wane of sun spots and faculae are due to planetary influence. The greatest of the world's weather forecasters goes by the planet's positions. I go by sunspots and faculae. Our dates always agree. The planets and sunspots are indissolubly connected.

A similar report was made by Prof. T. J. See of the United States Naval Observatory at Mare Island, Vallejo, California:

Great sun spots were noted at 4 p.m. today. The largest spot was slightly northeast of the sun centre and was made up of three parts roughly joined together by bridging, such as characterizes complicated spots.

This seems to indicate that the disturbance is destined to last several days. It appears to be increasing in magnitude and to be of the vortex or whirlpool type which often gives rise to disturbances of the earth's magnetism. The sun spots will reach the sun's central meridian in another day or two, and if any disturbance of the earth's magnetism is to occur from this cause it is likely to come about Friday.

The spots now seen are not believed to have any connection with the comet. If any aurorae or electrical effects occur, we should probably ascribe them to the comet rather than to the spots on the sun.

Professor Jacoby of Columbia disagreed. He said that they might be due to the passage of the comet's head across the sun which had occurred several days before.

In all, thirty sun spots were detected, indicating violent solar eruptions. The next day saw further unexpected developments:

ODD APPARITION
CROSSES FACE OF
AFTERNOON SUN

ASTRONOMERS STAND BEWILDERED BY
STARTLING WRAITH
OF THE SKY

MYSTERIOUS BANDS OF LIGHT CHALLENGE
ATTENTION OF VIGILANT OBSERVERS
WHO ARE PUZZLED BY SIGHT

Following close upon the wholly unexpected astronomical conditions that prevailed the preceding day, astronomers were further bewildered by an apparition across the sun's face at noon on May 19. A broad spectrum of light extended across and a considerable distance to each side of the sun. Professor Frost of Yerkes Observatory, who first sighted the phenomenon, said he had never witnessed its like. Hoping to get an accurate check on the strange spectrum, Professor Frost telephoned observers within a radius of 100 miles, notifying them of the peculiarity. The apparition lasted for less than half an hour.

A Lunar Eclipse

To complement the view of Halley's comet, there was a second celestial attraction on the night of May 23rd and morning of May 24 when the moon went into total eclipse. Many comet parties took in both events and had a pleasant evening because there was an intermission of only twenty-two minutes between the setting of the comet and the beginning of the eclipse.

The moon presented the rare phenomenon of remaining dimly visible while being in full shadow of the earth. A visible moon in eclipse was possible because sunlight bent around the earth and shone on the satellite in spite of the fact that the earth was between the two objects.

The eclipse began at 10:46 p.m. when the moon entered the earth's shadow. At 12:09 a.m. it was total. The unusual condition lasted an hour. At one o'clock the moon began to emerge, and at 2:23 a.m. the eclipse was over.

* * *

In the history of astronomy there was probably no period so short during which so many unexpected, novel, and peculiar events were observed within our solar system. One can almost sense the excitement and awe which inspired the scientific community of the world during those wonderful days. And the data amassed must have led to great leaps of theory and understanding. The rare combination of the state of the science, the apparatus available, and the occurrences combined to display celestial fireworks.

5. HALLEY'S COMET, CELESTIAL PHOTOGRAPHY, AND THE HAWAIIAN ISLANDS EXPEDITION

Photography was not invented in time to record the 1835 apparition of Halley's comet. But by 1910, celestial photography was ably serving astronomy. In August 1908, the Astronomical and Astrophysical Society of America appointed a Comet Committee to consider how the Society should use the opportunity offered by the return of Halley's comet in order to add to the knowledge of comets. By November 1909, plans were announced to create a "continuous photographic record" of the comet's presence. The Committee issued a circular letter of suggestion for the observation of the comet:

> The close approach of the comet to the earth promises
> an unusual opportunity for a study of the physical conditions
> that obtain in such a body and, as an indispensable basis for
> such study, the Committee recommends a photographic cam-
> paign for the production of a permanent record of the comet's
> appearance, as long and as nearly continuous as possible. The

comet's close proximity to the sun's direction at the time of
maximum brilliance imposes serious limitations upon this programme and widely extended cooperation will be required
throughout the whole circuit of the earth if this ideal of a continuous photographic record is to be even remotely realized.

Photographs taken in the Pacific were particularly important because two major events were going to occur while the comet was in view of the Pacific area. Halley's comet was to transit the sun, that is, pass in the line of view between the earth and the sun on May 18th. And the tail of the comet was to brush by the earth that same evening, the Pacific being in direct line with the tail.

Observing Shelter at Diamond Head.

The Hawaiian Islands expedition was the Committee's major effort. The National Academy of Sciences funded the project. Mr. Ferdinand Ellerman of the Mount Wilson Solar Observatory went to Honolulu in April 1910, to establish a temporary observatory on the island of Oahu. The site Ellerman picked was on the south slope of the world-famous Diamond Head, an extinct crater near the island's southeast end. Though the crater's rim is very narrow with steep slopes and deeply furrowed ravines, Ellerman located a fairly level bench about one hundred and fifty feet above the ocean and three hundred feet back from it. Strong winds blow almost constantly at that time of the year. The wind eddied around the ridges of the crater walls so it was necessary to anchor the shelter which was built of light framing covered with canvas. The observatory roof rolled on casters running in guided rails. During exposures Ellerman often had to cap the lens until the telescope could be steadied after severe gusts of wind. None-theless, a large number of excellent photographs were obtained.

The site proved to be the best place on the island for observing Halley's comet. "Time and again, during the morning appearance of the comet, photo-graphs were taken with a clear sky while three or four miles northwest towards Honolulu it would be raining."

On May 18th Ellerman carefully observed the sun's face with the 6.4 inch visual telescope for any traces of the comet in transit. Viewing was good enough for the granulations of the solar surface to be visible, but nothing of the comet or its nucleus was seen. It was calculated that only a nucleus of ten miles in diameter or larger would be visible; a smaller nucleus could not be detected even through the high-power telescopes used on Oahu. That same evening the comet passed closest to the earth, also in clear view of the Pacific.

Many observatories produced photographs of the comet of a quality that were judged to be "of conspicuous technical excellence." Photographs were reproduced and catalogued by Lick Observatory. The American Astronomical and Astrophysical Society published forty-seven photographs taken at Diamond Head with notes on the photographs as part of the Comet Committee's report for 1909-1913. With these publications the scientific community realized its ideal of producing a permanent continuous record of the comet's appearance.

6. UP, UP, AND AWAY:
THE COMET OBSERVED FROM BALLOONS

Balloon ascents were made by astronomers from several places in America and Europe to observe the comet through thin atmosphere and collect samples of air which might contain gases from the comet's tail. Several of these ascents were more on the order of ballooning adventures.

AN ASTRONOMICAL OBSERVATORY IN THE UPPER AIR: Taking observations of Halley's comet from a balloon.

The helium balloon and Halley's comet were depicted in the drawing "An Astronomical Observatory in the Upper Air: Taking Observations of Halley's Comet from a Balloon." Three observers watched the comet from the balloon's basket. A vastly oversized drawing of the comet's nucleus and part of its tail are the featured subject of Henri Lanos' drawing. Illustrations such as this one led people to be disappointed with the comet's actual appearance.

* * *

Wall Street broker Roswell C. Tripp, ex-Yale athlete and member of several All-American football teams, went from Grand Central Station with a few old college friends to Springfield, Massachusetts, where he chartered the balloon *Springfield* to see the comet. While in Europe the year before, Mr. Tripp made several successful flights and met the pilot Mr. Arnold, winner of the International Race held in Vienna. Arnold managed the ascent. Before boarding the train, one of the party said, "Mr. Tripp is the only one of us who has ever *seen* a balloon, but we have been training in the Harvard Club gymnasium for the last week. We will be equipped with photographic and telescopic instruments, and intend to make this a scientific as well as a novel adventure."

The expedition hoped to reach a height of at least 10,000 feet and intended to stay up as long as practical. One of the men carried a bag of salt to sprinkle on the comet's tail.

* * *

Aeronauts A. Holland Forbes of Bridgeport, Connecticut, Vice President of the Aero Club of America and holder of the Lahm Cup, and James Carrington Yates, astronomer from New York, went aloft to photograph Halley's comet through the thin atmosphere at 18,000 feet. They returned with valuable photographs and bad bruises. The thrilling flight began at Quincy, Illinois, Monday evening, May 10.

The day after the two aeronauts lost control of the balloon *Viking* and landed at Horse Cave, Kentucky, the headlines in the *Seattle Post-Intelligencer* sketched the story:

AERONAUT TELLS OF FEARFUL FALL

Forbes, Freezing Above the Clouds,
Ripped Bag as Last Resort for Descent

Mr. Forbes:

> Tuesday morning we encountered intense cold and a severe storm at 16,000 feet. That afternoon we ran into another one. Shortly after we shot up to 20,000 feet, too high for safety. We were descending, but letting gas out by the valve did not bring us to the ground as fast as I wanted. I used the rip cord before we lost consciousness entirely. The rip cord didn't work the way it is supposed to. Instead, it ripped the bag open almost from top to bottom. The descent was terrific. When we were about 800 feet from the ground I saw men below me in a plowed field. I tried to tilt up the basket so Mr. Yates and I could spring for the rigging when we got close to the ground. For the last hundred feet there was very little gas left in the bag. It fell like a stone. The basket landed bottom squarely down under the weight of the bag. The rubber air mattress we carried on the basket floor prevented us from breaking our backs.

Both men were badly bruised and the balloon partly wrecked. By the next morning Forbes and Yates were resting comfortably and by the end of the week they were able to journey to their homes in New York.

* * *

Astronomer and University of Pennsylvania Observatory Director Professor Charles L. Doolittle twice missed seeing the comet. He had been invited to accompany a crew from the Philadelphia Aeronautical Recreation Society in the balloon *Philadelphia II* on the night of May 20 hoping to reach a high altitude and see Halley's comet. Clouds prevented observations from being made, and the balloon landed in Milford, Delaware. The next night, May 21, Professor Doolittle ascended again, accompanied by Dr. Thomas E. Eldridge, Dr. George K. Simmermann, and Detective A. L. Millard of the Standard Secret Service Bureau. The balloon rose at 6 o'clock in the evening, passed over Califon, New Jersey, and landed in an open field three miles from there at Crestmoor. After the landing they were brought in, balloon and all, by Alden Silker, a farmer. The party reached an altitude of 6,500 feet, but even at that height the clouds were so dense that they did not see Halley's comet.

* * *

Typical were the reported results and findings of Professor George O. James of Washington University. The balloon *St. Louis III* left St. Louis at 6:30 on the evening of May 18 to explore the comet's tail. Professor James led the expedition, with Captain John Berry, a veteran pilot, as the guide. Reporter Andrew Drew accompanied them. The balloon belonging to the St. Louis Aero Club was placed at the expedition's disposal.

The astronomer intended to make observations during the transit of Halley's comet across the face of the sun. The purpose of the expedition was to determine, if possible, what effect the passage of the earth through a comet's tail has on atmospheric conditions and to observe possible meteorite showers. "The most I expect in the way of positive results," said Professor James before ascending, "is a faint auroral display or a hazy illumination spread over the heavens."

The equipment in the balloon consisted of an aneroid barometer, a sensitive thermometer, and a finder for his telescope. Enough ballast for remaining up all night if necessary was carried, but a landing about midnight was planned. The balloon took a northeasterly course in a light breeze and soon was out of sight. It was sighted at St. Charles, Missouri, twenty miles from St. Louis, at 7:10 o'clock. The balloon traveled sixty-five miles, landing in a wheat field near Hillview, Greene County, Illinois, at 11:20 o'clock. Professor James reported, "I noticed no phenomena which I could confidently attribute to the effect of the passing of the comet's tail. There was no faint phosphorescence in the skies, no auroral display, and no showers of meteors within our vision such as many astronomers had predicted."

Also in St. Louis, Father Martin Brennan of Kenrick Seminary was invited to ascend in a balloon with other astronomers to prove that they would not be killed by the comet's tail and to collect experimental data. Father Brennan declined, not because of any fear he had of the comet's tail, but because he "did not take kindly" to balloons.

* * *

The most interesting experiment was carried out in Germany. The balloon *Abercron* was the only one of a large number of balloons making ascensions which succeeded in observing any unexpected activity. Reaching an altitude of 11,385 feet, the aeronauts noted extraordinarily brilliant horizontal lights like meteors. They filled four bottles with air at that height to be examined for the presence of meteor dust and other celestial ingredients. The aeronauts were finally forced to land because of an electrical storm.

From thirty-one places in the world sonder balloons were released to determine any effects in the upper atmosphere ten miles above the ground caused by the earth's passage through the comet's tail. Balloons were sent up from Omaha, Nebraska, and Blue Hill Observatory, Hyde Park in Pittsfield, Massachusetts. Among the European cities where the experiment was performed were Manchester, England; Berlin and Hamburg, Germany; and Strasbourg, France. These sonder balloons, which were six inches in diameter, carried instruments to record altitude and temperature, and a parachute to which the delicate instruments were attached. When the balloon filled with hydrogen gas exploded, the parachute brought them safely to earth. There was a message and directions in the basket stating that the finder would receive a reward for sending the instruments to the observatory. The experiment was repeated at each location on each day through May 21.

7. THE COMET, WIRELESS TELEGRAPHY, AND THE BREAKTHROUGH TO MODERN RADIO

The United States Hydrographic Office issued a special bulletin calling on all mariners to make careful notes of unusual meteorological conditions which might be caused by Halley's comet. The bulletin said:

> It is requested that all ship masters who view the comet will inform the United States Hydrographic Office of the brightness of the object as compared with bright stars in the heavens, the angular length of tail, comparing the length with the angular distance between bright stars near it, its form, and color. To make these observations it is suggested that ship masters use their binoculars, spyglasses, or the low power eyepiece of their sextants.

Mariners were to make careful observations for variations of the needle, which might have been affected by the comet. They were to keep close watch for unusual deflections, noting their exact time, amount, whether periodical, and any other noteworthy features.

The maximum effect was expected between the 16th and 20th of May, particularly on May 18. Wireless operators were to pay particular attention to static effects during this period, and to note anything unusual. The operator hearing unusual noises in the telephone noted the time, whether there were any meteorites observed, the ship's head, and direction of the antennae. The meteorites were to be logged by the watch officer with details of time, size, direction, and duration.

Many amateur wireless plants were scattered about the country, besides those of regularly organized companies doing a commercial business. All operators of land based plants worked in conjunction with the Hydrographic Office, keeping the same records requested of operators on ships. Scientists generally believed there would be little or no effect noted in wireless transmissions. Professor O. C. Wendell of the Harvard Observatory said he did not expect any interference in wire or wireless communication.

Halley's comet did not interfere in the slightest with the working of wireless telegraphy up to May 13. On that day local stations reported the wireless worked better than usual the night before and all day. Nine ships reported their positions, and one of them, the big liner *Manchuria,* was far across the Pacific. But no one attributed the improvement to the comet.

However, the officers on board the steamship *Antilles* reported two curious phenomena occurring on May 13. There was a continuous loud ticking heard through the phones attached to the wireless instrument, and a shower of meteors was seen in the southeast.

While many watches had been kept on ships to detect magnetic disturbances caused by Halley's comet, the first report of one made to the United States Hydrographic Office was from Capt. Jones of the steamer *Idaho* of Hull, England. Before sailing from Hull the captain was asked to look out for his compasses on May 16. A careful watch was begun from midnight of May 15 onward. Before dawn a brief sighting of the comet was made, but clouds soon hid it from view. At 8 a.m., while the ship was in mid-ocean, the compasses swung gradually reading the maximum of eight degrees in a few minutes. The compasses worked slowly back to normal in the next two days.

Finally the Hydrographic Office's project reached an almost ironic conclusion. *The New York Times* headline tells the story:

TOO MANY COMET REPORTS
Navigators' Zeal Wiring Observations Overburdens
Hydrographer's Office

On May 23 Captain Knapp, Chief Hydrographer, announced that he obtained more information about the comet than he could use. Reports as requested came in such numbers that Captain Knapp foresaw the speedy exhaustion of the limited sum of money allowed annually by Congress for telegraph and cable expenses of the office. They all came "collect" and the hydrographer could devise no way to stop them.

In the long run, of course, the information received was extremely valuable. On behalf of science and knowledge it was wise for the United States Hydrographic Office to carry out this project. It is equally rewarding to find that ships at sea and their captains and crews responded wholeheartedly.

* * *

On May 17, Dr. Lee DeForest, prolific inventor, introduced his Audion tube— a third electrode added to the two-electrode electron tube or diode—creating the triode. The triode *amplified radio signals and generated oscillations,* two functions which separate the wireless and crystal radio from radio as we know it. Dr. DeForest installed a receiver and aerial equipment on the roof of the Arcade Annex in Seattle, to demonstrate the comet's effect on his new wireless apparatus. During the morning and afternoon of the 18th a sound of roaring and crackling and hissing was detected through Dr. DeForest's new receivers. He attributed these disturbances to electrical discharges caused by Halley's comet. The audion, Dr. DeForest explained, is far more sensitive than the spark wireless telegraph.

The results of DeForest's demonstration showed that the audion was able to pick up and amplify electronic and electromagnetic signals which went undetected by the wireless equipment using the diode. His equipment added a new dimension to radio's capability.

This new tube also contributed to a marvelous succession of events in the history of astronomy. The audion allowed the astronomer to explore space through sound generated by electric and electromagnetic activity in space. So Halley's comet figured in demonstrating a breakthrough in radio reception, whereby the audion allowed the listener to hear what was formerly unheard.

8. HALLEY'S COMET AND THE TELESCOPE TRADE

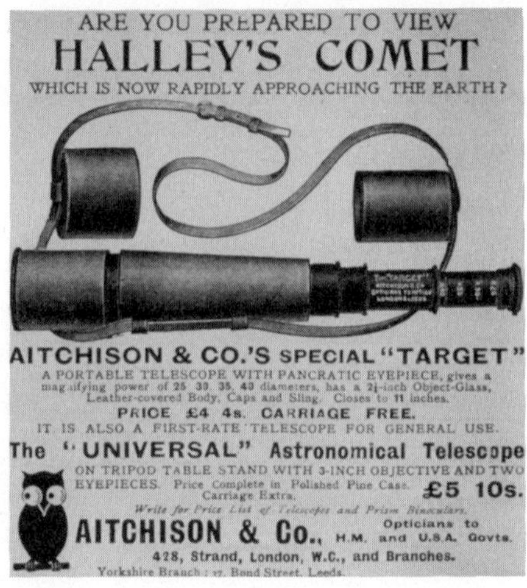

Halley's comet had the whole telescope, spyglass, marine, and field glass business booming at a rate the trade had not known since the Civil War. Business in the week before May 18 was the best in years, and the next week was expected to be even better.

People of all classes became astronomers. They were buying first class telescopes that sold from $50 to $250. The most popular telescopes were those by French and English makers. Both were mounted on tripods and fitted with precision lenses. The barrel of the English instrument was of polished brass, while the French telescope was covered with white leather. Through them Halley's comet was a sight of immense beauty.

People who could not afford a telescope bought smaller instruments. Marine, field, night, and opera glasses sold from a few dollars up. The cheaper glasses sold in great quantities in pawnshops. One day a shop on Third Avenue in New York City filled a window with glasses selling for one dollar up to ten dollars and the next day all were sold. A Broadway dealer said, "We sold more in the past three months than we sold between the end of the Civil War and the start of 1910. That's forty-five years worth of business in three months!"

The wholesale dealers of the Maiden Lane district exhausted their supplies. The story was the same all over the country, all over the world. Every telescope and every set of binoculars, spyglass, surveyor's and opera glass hidden away in trunks and drawers must have been dusted off for use and pointed to the sky to salute the celestial marauder who came, it seemed to many, with motive. Some saw him as a pirate—especially the spyglass types—others, as the wrath of God come to smite us with death and destruction.

Retailers raised prices twenty per cent and people paid gladly. Everyone—and everyone everywhere on earth—wanted to see Halley's comet. Importers tried to buy back instruments from local retail stores and from other parts of the country, but nobody had any left.

Downtown hardware stores were selling telescopes manufactured expressly for the comet trade for a dollar. The barrel was made of pasteboard in sections and according to the label on the side:

HALLEY'S COMET NOVELTY TELESCOPE
Will Present a Beautiful
Picture of the COMET

Peddlers, who sold whatever happened to be the fad, said the last week before the comet came closest to the earth would see many Halley's comet novelties on the market. They hawked kerchiefs and shawls with the image of the comet painted on, costume brooches and hat pins in the comet's shape, postcards depicting the destruction of the world, pamphlets on the comet's history, charts showing the exact location in the sky each night.

Then it all stopped. Nobody bought telescopes anymore. The novelties, along with the pasteboard telescopes, found their way to the trash, and everything worth keeping was stored away until 1914, when the trade was called upon to manufacture field glasses for the armies and navies of the world. From 1914 to 1917, the years of World War I, the optics trade once again profitted, this time from the folly of mankind on earth, while the last time, 1910, it had profitted from mankind's interest in the heavens.

9. INTERESTING SCIENTIFIC EXPERIMENTS

Naturally a number of experiments were proposed and carried out. Among the most interesting were the following:

An American astronomer conceived the idea of weighing the comet by the deviation it produced in the earth's orbit. Noting this plan, *The Scientific American Supplement* said, "We can hardly believe that the effect produced will be one one-thousandth part of the least measurable quantity…." Nonetheless, the commentator, P. H. Cowell, went on to remark, "…The speculation is most interesting in view of the fact that there are unexplained phenomena in planetary movement."

Professor Birkeland of Christiana, Norway, a well-known observer of magnetic activity, conducted magnetic and meteorological studies in northern Norway from May 7th to June 1st. The professor hypothesized that the comet's tail consisted mainly of "electrical corpuscular rays." He expected the magnetic forces of the earth to attract the rays in the Polar regions causing visual displays in the auroral zones.

Several experiments were planned to take place on May 18th. It was suggested that samples of the atmosphere be bottled up on the night of May 18th to see if it contained anything unusual. *The British Mechanic and World of Science* observed, "A cellar full of bottled comet would be a very nice legacy to hand down to posterity."

On a more serious note, an experiment described in the *Bulletin de la Societe Astronomique de France* "... suggested the liquefaction of a large quantity of air which could afterward be treated by fractional distillation..." and cometary matter be captured for study. The scientist who proposed the experiment, C. E. Guillaume, pointed out that minute quantities of rare gases like krypton and argon are extracted in this way from immense volumes of air. *Scientific American* commented on the experiment saying it was unfortunate that the chance to capture a sample of the tail was not taken. "The passage of the earth through a comet's tail is so rare an occurrence that no opportunity should be missed" to study it.

But, in fact, the experiment *was* performed, a rare instance of a *chemical* study of the comet carried out as a by-product of an industrial process. Acting upon Guillaume's suggestion, Georges Claude attempted to trace cometary matter among the residual inert gases left from the liquefaction of air. With his apparatus at Boulogne-sur-Seine he was able to treat 350,000 litres of air per hour, and to detect the presence of one-millionth part of any extraordinary gas. Experiments carried out on May 17 (4 hours), 19 (9 h to 12 h, Paris M.T.) 20, and 23 failed to reveal any difference of density in the residue that would be greater than the probable error of observation.

A tower was built on the peak of Mt. Wilson near Pasadena, California. Attached to the tower were metal plates coated with glycerine to catch dust particles from the comet's tail, but no particles were discovered. The astronomers found nothing to indicate that ordinary conditions had been affected.

The absence of positive results in all cases stems from the same cause; the amount of cometary matter making its way through the thick atmosphere was too slight to leave any measurable trace.

II
Three Astronomers

10. ASTRONOMER MARY PROCTOR: "THE LONELY WATCHER ON THE TOWER"

The appearance of Halley's comet in 1910 created eager audiences suddenly fascinated by astronomy. Three hundred and fifty astronomers in the United States, and 1,000 worldwide, trained their telescopes on the comet. Eminent astronomers lectured widely, wrote articles for journals such as *Scientific American* and *Astronomy,* intellectual reviews like *Current Literature* and *American Review of Reviews,* and popular magazines such as *Collier's* and *Harper's.* Even the finest children's magazine of the day, *St. Nicholas,* engaged a prominent astronomer to explain the astronomical facts about the comet to America's children.

Some individual observations and accounts were particularly wonderful. Mary Proctor, an outstanding amateur astonomer herself and daughter of a highly-regarded astronomer, was especially interested in comets. Halley's comet's arrival gave her an opportunity to write for the largest audience an astronomer was likely ever to have up to that time, the vast readership of *The New York Times.* From April 26th through the end of May, her daily reports of her observations of the comet appeared. The readers saw the astronomers' predictions unfold and heard Miss Proctor describe what the readers themselves had observed each night.

Long before Halley's comet was visible to the naked eye, Mary Proctor saw it. On August 29, 1909, the comet was recovered by Max Wolf, celestial photographer, at Koningstuh Observatory; Mary arrived three weeks later at the Williams Bay, Wisconsin home of professor and Mrs. E. E. Barnard, to stay as their house guest. The night of September 16th the great Yerkes Observatory refractor telescope with its 40-inch lens was in the care of Professor Barnard. He invited Miss Proctor and his niece, Miss Calvert, to come to the observatory at 3:00 a.m. The early morning of September 17, 1909, was one of the most eventful times in Mary Proctor's life.

> Making a first visit to the observatory in the darkness preceding dawn was an experience in itself, but the glimpse of the comet after its absence of seventy-five years is one never to be forgotten, nor is it easy to describe. For the first second or so, all seemed darkness as I gazed down the length of the great tube (63½ feet) into the opening beyond. I saw nothing, and an intense feeling of disappointment overwhelmed me as I realized and stated this fact, but Professor Barnard remarked in his whimsical way: "Surely you did not expect to see the comet with a tail?" Then he advised me to keep on looking, and even while he spoke I saw a faint, very misty outline. "Is it exactly in the center of the field of view?" queried Professor Barnard when I told him that I had seen a nebulous-looking object, and when I replied in the negative, he informed me that that faint object I was looking at *was the comet,* which eight months later I saw in all its splendor from the tower at the top of the *Times* building in New York City.

By April 26th, 1910 the comet had come within range of naked eye observation. Miss Proctor was in New York City at the time. Despite the smoke and electric lights of the city, she very much wanted to see the comet. While walking along Broadway that afternoon, and glancing in the direction of the *Times* building, New York's tallest structure at that time, it occurred to her that this would be the ideal location for observing the comet. She entered the building and made her way to the office of Mr. Van Anda, assistant editor of the *Times,* and explained what a desirable spot the summit of the *Times* building would be. Mr. Van Anda wrote a note giving her permission to ascend to the tower each night as long as the comet was visible to the naked eye. "The next morning," Miss Proctor wrote, "it was indeed a case of 'Call me early, Mother, Dear,' but an alarm clock served the purpose equally well on this momentous occasion."

She wrote further:

> Promptly at three o'clock the permit was presented to the janitor, and the writer, ascending in the lift, was transported to the twenty-third story, and escorted up a spiral staircase leading to the tower. The guard unlocked the door and the writer, stepping out on to the parapet surrounding the tower, gazed eastward for the comet, which failed to materialize, owing to a dense haze. Awaiting until dawn, the idea of seeing the comet was given up.

Miss Proctor's report on the sighting appeared as a letter to the editor on April 29th. She presented details of her observations in layman's language, retaining the substance and flavor of serious astronomical study. From the outset she demonstrated an ability to artfully weave observational facts, astronomical data, and lyrical descriptions of what she saw.

> Then came May 4th, a bitterly cold morning; but the stars shone brightly and there was every hope of the comet being visible from the tower heights. Calling down to the janitor to make known the good news, the balcony was soon filled with eager members of the *Times* staff, who were thus enabled to obtain a view of the comet. By means of a field-glass thoughtfully provided by Mr. Van Anda, it was possible to see a further extension of the train, making it in all thirty degrees in length. Spurts of light like tiny waves seemed to flow out from the nucleus to a distance of two or three degrees. At twenty minutes to four, the writer on looking downward at the horizon, was startled by what appeared to be a streak of flame, but as it rose higher it appeared to be the crescent moon, which with the comet and the planet Venus, completed a wonderful trio. The comet remained visible, resembling a bright star with a slender stream of silvery mist trailing a few degrees after. By four o'clock it had faded in the light of approaching dawn.

The Romance of the Comets

Beginning with the May 7th issue of *The Times,* Miss Proctor's reports appeared under her by-line rather than as letters to the editor.

The comet was obscured by clouds in New York City for several nights.

It was not until the morning of Friday, May 13 that the comet once more deigned to reveal itself to the straining eyes of the lonely watcher on the tower.

The New York Times

Like a poet of science Miss Proctor wrote:

> The nucleus resembled a golden globe immersed in folds of gauze. Each moment it became more clearly defined, finally glowing as brightly as a star of the second magnitude.

The New York Times

On the crucial night of May 18th:

> The parapet surrounding the tower was crowded to its utmost capacity by a favored few awaiting they knew not what, for a report had gone forth that we were scheduled to pass through the train of the comet. Below us we could see comet parties in progress on the roof gardens of some of the leading hotels. Sounds of merriment occasionally reached us, but by half past ten we—that is, Miss L., who had offered to share the lonely vigil with the writer until dawn—were the only watchers on the tower. The hush of a great silence had gradually fallen over the city, and in silence, too, we watched the eastern sky for any further trace of the comet.

The Romance of the Comets

What Miss Proctor and her companion stayed to watch was *not* supposed to happen, according to professional astronomers. Having brushed past the earth on May 18th, the comet would be visible in the west in the evenings just after sunset instead of in the mornings just before sunrise in the eastern sky.

Believing that the comet's immensely long tail would appear that morning in the east, Miss Proctor decided to watch through the night. She and her companion saw many vivid displays in the sky associated with the comet.

> 11:10 red flash (auroral); 11:22 flash resembling an arch of glowing white surmounted by a crest of crimson ... At 2 o'clock a meteor flashed across the eastern sky, downward in the direction of the star Gamma in the constellation Pegasus. It was bluish-green in color, pear-shaped in appearance ... The display was vivid while it lasted.

The Romance of the Comets

Then at 2:45 a.m. Mary Proctor observed streamers "reaching from the eastern horizon, below Gamma Pegasi, and curving upward through Aquarius as far as Altair, and brighter in appearance than the Milky Way." The path of this band of light was along the same path where the comet and its tail had been seen the night before. She recorded all the pertinent data to confirm her observation. She also sketched it, as astronomers are trained to do.

When the report of her sighting appeared in *The Times* the next day, an astronomer replied with a statement published in an afternoon paper: "Someone thinks she saw the comet in the eastern sky, when it is really in the west." If in error, Miss Proctor would be in monumental error, off in her sighting by twelve hours and an entire horizon. Fortunately, the same streamers were observed by other astromomers and her report was confirmed. Miss Proctor wrote:

> One can imagine the anxious time experienced while awaiting confirmation of the observation, but it came in due course from Yerkes, Lick, Argentine Republic, South Africa....
>
> *The Romance of the Comets*

Her friend, Professor Barnard, was among those astronomers who saw what she described as "a narrow twilight...which seemed to extend along the eastern horizon..."

A few days later a fellow-amateur astronomer wrote to *The Times* commending Miss Proctor on her success. M.H.R. of Daytona, Florida, wrote:

> To the Editor of *The New York Times:*
> Having more than average interest in astronomy, because of considerable study for a score of years and the possession of a twelve inch telescope, it gives me pleasure to express hearty appreciation of the achievement of Miss Mary B. Proctor in "locating" the tail of Halley's comet by looking to the east while all the professors were gazing in the opposite direction on the night when the predictions were that the earth would be bathed in comet dust.

To place Miss Proctor's daring in perspective, he recounts an incident in one of Sylvanna Cobb's stories:

> ...which some of the old boys like myself may remember, when some people in an early age of the world decided to choose as King the man who should first see the sun in the morning when the tall candidates were standing on a level plain. One wise man, not so tall as the others, turned his back to the east and looked intently toward the west, and the others smiled at this seeming indifference. But, when the sun, not yet risen, was gilding a mountain top to the west the man shouted "Look! Behold the sun!" And when the whole assembly turned to look at the sunlight he suddenly turned to the east just in time to see the sun when it was appearing above the horizon. And they made him King. So Miss Proctor discovered the tail of the comet while the most famous professors were looking in the other direction. The discovery is worthy of commendation.

A *Times'* article by Mary Proctor carried ominous headlines and a prediction which we may see fulfilled in 1986:

COMET'S VAGARIES
MAY MARK ITS END
____

Evidence That It Is Breaking
Up, Similar to Those of Others
That Have Disappeared

Twin Comet May Appear in 1986

She wrote, "Is a celestial tragedy being enacted...? Startling rumors...have been circulating to the effect that...the nucleus of the comet has split in twain." After recounting the division of Biela's comet observed in 1852, Miss Proctor reported, "The discovery made by Dr. A. E. Douglas of the University of Tucson, Arizona. He states that the head or nucleus of Halley's comet has divided into two parts."

Miss Proctor reminds us that on May 19th two fragmentary streamers or tails were observed; this condition is usually a sign of a comet breaking apart. Impressing her name upon the future, upon 1986, and staking her claim to a prediction, Mary Proctor wrote:

> On the evening of May 25th...four more meteors flashed into view....Were these meteors the celestial heralds announcing a future display seventy-five years hence of...the scattered fragments of Halley's comet...?
>
> This may be too early a date for the final disintegration and at the next return in or about 1986 twin comets may greet the eyes of astronomers, but it is only a matter of time when the comet or comets will finally go to pieces. One feels as safe in making this prediction as Halley did when he called upon candid posterity to acknowledge that his prediction of the return of Halley's comet in 1753 had come true. The writer as confidently predicts that if Halley's comet continues in its present course of dividing in two and losing its tail and apparently scattering celestial fire rockets along its track it will doubtless not survive more than two or three centuries at most. "Candid posterity" can alone verify this prediction which has the charm at present of being "beyond our ken."

Will we observe twin comets made from Halley's comet, or will the separation occur later in the three hundred years that Miss Proctor allows? And, which do we choose? Of course we all hope for a Halley's comet whole and hearty.

Mary Proctor contributed valuable reports to her readership; they were scientifically accurate, simple, clear, direct, factual, warm, poetic, enthusiastic. She represented the science well, gaining the admiration of an appreciative audience. And she reminds us of science at its best, an adventure of the mind and spirit, the quest for truth about the universe and our place within it.

11. FRENCH ASTRONOMER CAMILLE FLAMMARION SAID "WE MAY DIE LAUGHING WHEN THE COMET COMES"

Los Angeles Times Sunday Magazine

The public of 1910 came to be familiar with the name and predictions of the French astronomer, Camille Flammarion. His opinions and predictions reported in *The New York Times, Los Angeles Times Sunday Magazine,* the French pictorial magazine *L'Illustration,* and *Busy Man's Magazine* were contrary to the predictions of the orthodox scientific community, particularly on the subject of cyanogen in the comet's tail. Flammarion claimed there might be enough to bring animal life on earth to an end. His astronomical studies at his private Juvisy Observatory were far less interesting to the world awaiting the earth's passage through the comet's tail than the images of fictional terror Flammarion created before the mind's eye. His comments on Halley's comet for the benefit of the press, either artfully or crudely planned, alternated between scientific and subtly cued images of the end of life on earth. This alternation of voices or intentions is heard again and again. In an article entitled "Halley's Comet and the Fate of Humanity: Some Interesting Prognostications of What May Happen Should the Earth Run into the Tail of Halley's Comet and Become Enveloped in its Poisonous Gases" Flammarion wrote,

> *The poisoning of humanity by deleterious gases is improb-
> able. Doubtless if* (italics mine) the oxygen of the atmosphere
> combined with the hydrogen of the comet's tail it would mean
> universal death with short shrift.

Surely "The poisoning of humanity by deleterious gases..." grips the mind far more powerfully than the weak near-negation, "...is improbable." Similarly "...universal death with short shrift..." stirs the imagination deeply, while the conditional "Doubtless if..." fails to animate the reasoning faculty sufficiently to overcome the fear of impending death.

The next passage Flammarion wrote "sounds" scientific:

> Anxious minds have, however, no reason to be tormented
> —uselessly, too—by these prognostications. Comet tails, it is
> true, are immense, but they are so light, so rarified, that the
> terrestrial atmosphere is like lead in comparison. (And now Flam-
> marion terrorizes again): "Even were our globe *completely
> plunged* into such a tail, we would, *without doubt,* be saved from
> a *cataclysm* by the atmospheric curtain which surrounds us."
> (Italics mine)

Flammarion was a clever rhetorician. "Lead" becomes "a curtain." Science becomes the certain voice of doom.

Flammarion also predicted that if the supply of nitrogen were reduced, "...the brain of every one of us would experience an unexpected sensation of physical activity and the human race would come to a sudden end in a paroxysm of joy, universal delirium and madness." He claimed also that if the tail contained carbonic oxide there would be a "...universal poisoning of the lungs."

Needless to say, scientific astronomers were outraged by Flammarion's fear-mongering. Like the Mercury Theatre's famous "War of the Worlds" Hallow-een radio prank in 1938, Flammarion's pronouncements stirred up much fear and ill-feeling. He was severely and justly criticized in the scientific press. In a letter to the editor of *The English Mechanic and World of Science,* Felix de Roy eloquently chastised Flammarion for predicting the destruction of animal life on earth. Speaking of the widespread disappointment over the comet's appearance, de Roy wrote:

> The real signification of the astronomical event we now have the pleasure to witness has been completely and very badly denaturated in the public's spirit here, under the influence of a small band of pseudo-scientists living in France. As soon as the fact of a probable transit of the earth through the tail had been announced, a dreadful story of "poisoned cyanide gases" went on, and expanded with a wonderful ease. This "fin du monde" story had found very firm believers....
>
> What I wish to point out is ... that the fears of the comet were *not* aroused by this celestial body *itself,* but by "a tale of how certain Frankish astronomers had foretold the destruction of the earth." When one considers that the publication of this tale induced many people in France, Spain, Italy, Hungary, etc., to commit suicide, may the author of this story not be considered as a public malefactor? I do not wish to search out here the real author of the tale; but those interested in the matter may conveniently find it in the "Bulletin de la Societe Astronomique de France", *"L'Illustration",* and the *New York Herald* (Continental Edition).
>
> In conclusion, may I reproduce here a statement of Mr. E. M. Antoniadi ("E.M." LXXVII. P. 299), who wrote in 1903: "It is, therefore, the duty of scientific students to defend the banner of truth and to pave the way for the general intellectual progress by facilitating, not hampering, the gropings of mankind." Does not this sentence apply to the present case?

Flammarion seemed to write and speak science, while writing and speaking science fiction. Even when he spoke more scientifically his remarks were colored with the language of the fantastic. About some of the facts regarding the comet, he wrote:

> Bathing in the effluvia of the electric, calorific, luminous radiation of the sun, it becomes impregnated with its rays, undergoing in its whole being fantastic transformations which lend it prodigious glory, develop it by multiplying, ten times a hundred times, its volume, lengthening it to millions and millions of kilometres by a kind of phosphorescence which always is extended away from the sun and gives rise to the formidable tails which filled with terror the souls of our ancestors.
>
> Thenceforth the wanderer's path takes it away from the ardent centre to sink into the deserts of immensity, gradually diminishing in size, becoming a sort of invisible bubble, and finally to find again the night of its aphelion in which for years and years it is lost to the eyes of astronomers on the earth. It goes away to a distance of five thousand million kilometres, into the ultra Neptunian night, in which its speed is gradually decreased to less than a kilometre per second. The total duration of its circuit is sixty-five years.

Flammarion had been working on his horror story for a long time. In 1893 and 1894 a story by him called "The End of the World" appeared in a scientific magazine, later being published in book form in Paris. The story tells of the destruction of the earth by a comet. It was translated from the French for the May, 1910 issue of *The Scrap Book* and was summarized in Edwin Emerson's scare pamphlet, *Comet Lore.* The last pages are a fine stirring of the sense of utter doom:

Already the comet had passed within the lunar orbit. Nearer and nearer it drew to the terrestrial atmosphere. It almost touched the hem of the earth's robe with its fiery fingers now.

Then the contact came. There was a flash on the horizon. A great sea of colored flame shot up. From a myriad throats came an agonized wail:

"The world is on fire! The world is on fire!"

It was the last sound that came from a human throat. Every man, woman, and child perished as swiftly as if a mighty wind had blown out their lives like so many candles.

A few seconds elapsed before the nucleus struck the earth. In those few seconds the atmosphere was ignited with a terrific explosion, and the earth was wrapped in a tossing, roaring, seething garment of flame. Rivers and lakes disappeared in hissing clouds of steam. Forests and cities flared up like matchwood. Then came the crash.

Like a great celestial projectile the solid nucleus of the comet pierced the egg-shell crust of the earth and buried itself in the semimolten interior. The comet tore its way on like a shot piercing the boiler of a battle-ship.

The earth was immediately converted into a planetary volcano. Oceans were spilled like thimbles of water, only to disappear in dense clouds of vapor which mingled with the fiery contents of the earth. Continents were twisted and torn like paper.

Rocks disappeared in that fearful heat and became like water. The earth seemed to reel as it was smitten by that mighty blow. A great rent appeared which extended from pole to pole, a planetary chasm out of which spouted geysers of molten rock, jets of flaming gas, great sprays of matter which were hurled out for miles into the blazing atmosphere, only to drop back in a great rain of fiery heat.

There was a vast indentation produced by the collision which was quickly inundated by this sea of fluid matter. Mountains disappeared like cakes of ice in pots of boiling water.

Once more the world was reduced to the ball of incandescent liquid and gas from which it had painfully evolved into a living globe through millions and millions of years. It shone like a little sun in the black canopy of the heavens, brilliantly self-luminous. For thousands of years, perhaps for millions, it was doomed to wander through space, a rock-bound, sea-swept world reduced to the pitiful condition of a globular furnace in which the beginning of a new life was being forged. Its dull-red glow for ages bore testimony for more fortunate planets to the frightful cataclysm through which it had passed.   *The Scrap Book*

Flammarion's personal behavior and character were of a piece with his horrific predictions. Who was this astronomer who read humanity's fate in astronomical facts? Born in 1842, the son of a plowman, "When...a boy...Flammarion set out for Paris to find fame. He rented a little room, attended classes, studied fourteen hours a day," and by 1858 he gained a post in the Paris Observatory.

By an account in the *Los Angeles Times,* Flammarion was "The greatest living astronomer of France." Against a photograph of Flammarion standing beside a globe showing figures from the zodiac, the *Los Angeles Times'* article disclosed these two anecdotes about the astronomer:

> Mme. Flammarion is devoted to her husband. She refuses
> to allow any barber to cut the complicated locks of her "Flamme"
> as she calls him. She cuts his hair herself and then...stuff(s)
> her footstools and furniture with it.

The second anecdote had it that a visitor to his observatory, a lady of "fair shoulders" wearing an evening gown, once stood between Flammarion's telescope and the astronomer. "Flammarion immediately permitted himself a gallant compliment upon this charming barrier between the stars and himself." The lady was so "gratified by the compliment" that she "left him her skin in her will to bind a book. A copy of one of his own works may be seen, bound in human skin, upon his table." Among the astronomers who came to the public's attention during the days of Halley's comet, Flammarion filled a unique niche. Hair-stuffed furniture and a book bound in human skin? The astronomer was portrayed as something like a wizard—some science, some magic, some eccentricity. Flammarion represented the ancient past: the scientist's ancestry associated with the art of reading signs in the heavens. He brought the art to its inevitable conclusion, having the celestial and the mundane meet in an apocalyptic cataclysm.

He endorsed a brand and line of excellent binoculars which appeared in the better French and Italian picture magazines during the months and weeks prior to the comet's appearance and sold widely.

At the time of Halley's appearance, M. Flammarion was portrayed as sinister, evil, and unethical. From today's point of view, Camille Flammarion, flamboyant astronomer, brought a touch of the Night, the old ways of seeing back to mind, recalling astronomy's ancestry and heritage.

12. HALLEY'S COMET AND THE MAGAZINE *FLAMING SWORD*

The Cellular Cosmogony

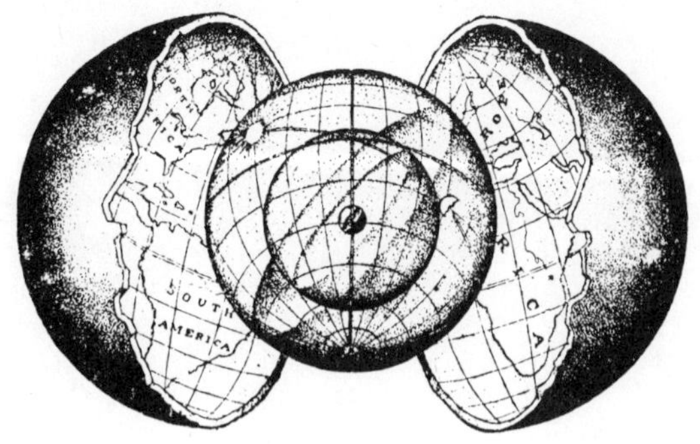

Discovered by Koresh
In 1870

As divergent as the opinions and points of view expressed regarding Halley's comet and the universe were in 1910, all were based upon the Copernican and Newtonian theories describing a solar system revolving about the sun and a vast universe beyond. Outside of this view in 1910 there was a radically different view, neither Copernican nor pre-Copernican, completely novel to systematic "study" of the cosmos. In the system of Koreshan Universology the earth is pictured as a living cell, the continent and oceans occupying the *inside* surface of the world instead of the outside. This unusual map is the fundamental image and foundation of Koresh's "revelation." Since 1886 this picture and principles and insights arising out of it had been discussed and described in the *Flaming Sword,* a beautiful magazine for the propagation of the discoveries of a Dr. See, known as Koresh after receiving a revelation.

Halley's comet and its appearance in 1910 was explained in *Flaming Sword* from the point of view of Koreshan Universology. The first article reported the failed announcement of the fatal night of May 18. The second strikes a blow against the Copernican system. These two essays or editorials read like Jorges Luis Borges stories, the form being that of an essay, while the content is the weaving of a story teller's fantasy. The essays express the devout beliefs of the writers and readers of the *Flaming Sword.*

The fact that the world continued unscathed by its brush with the comet was taken to confirm the Koreshan revelation in:

THE LOST TALE OF THE COMET

The Scare-Crow of Our
Modern Astronomers

The Lost Tale is about the tail of that comet, and how it
destroyed America. What an exodus has taken place! Here is
the great steamship loaded with stars from a New York opera

troup, including Caruso, all en route for Europe. Here is Signor Bonci hurrying after them, wild with excitement, with his hair flying to the winds: "Only save me, good captain, from being on this side of the water when America gets hit by that cyanogen bomb. Take me back to Italy, and take me back quick."

For three days the New York dailies announced that the fatal night had come. Either Halley's comet was to burst in a shower of meteorites capable of inundating the continent, or a deadly gas was to suffocate its inhabitants. They feted the comet like the New Year—in champagne. It was just *te moriturii salutamus.* The great restaurants were thronged. The mayors of small towns ordered the bells to be rung. Everywhere the same expectancy; everywhere the end of the world.

This fear and terror result from modern atheistical science, which predicts a distinct end of the physical universe. From childhood this fear is inbred. The greatest deliverer from these fears is an assured conviction that the God of Abraham, of Isaac, and of Jacob is not going to destroy this physical frame, the earth; that he has in these last days sent a revelation of himself, of his character, and his purposes toward mankind. The Cellular Cosmogony demonstrates that the physical universe is an eternal structure. The inhabitants of America have rejected this revelation of god, in preferring the old and effete system of Copernicus, which grew up in the Middle Ages. They are afraid, because they are living under the domain of fear, which is ignorance.

The comet has always been a source of terror to those who think in Copernican figures. The length of its tail is sufficiently amazing. It is supposed by these astronomers to measure more than one million miles across at the point where the earth (a little planet, swinging around at an enormous rate in space) darts through it. The head of the comet has been sighted at, supposedly, sixteen million miles from the earth. In rebuttal of all this testimony founded upon *hypotheses,* for these immense distances originate from *the primary guess that the earth is convex, the Cellular Cosmogony proves that men are dwellers in a cell only twenty-five thousand miles in circumference, with a diameter of eight thousand miles.* . . . . (Italics mine)

The earth is a cell and we live inside it! A hollow earth theory thriving in 1910!

Instead of any cause for apprehension at the sight of our celestial visitor, with all its apanage, one has only to remember what a comet is in terms of Koreshan Universology.

Now we get some unusual and unfamiliar language, which need not be translated, just enjoyed.

It is simply caused by the breaking in of zones of cruosic energy generated at the colure. These form lenses that reflect the sun's rays. Comets spirate about the sun—the central sun, not its visible focalization, and then plunge into it to feed its fires. Let people go to bed and sleep nights henceforth, when a comet is in view....

The essay concludes with a recapitulation of the themes introduced: the earth was not devastated by the comet or its tail and the terror in the hearts of many was brought about by the "brutal notion implanted from childhood" that the earth is a sphere located in the Copernican system.

A second essay in the *Flaming Sword,* "Halley's Comet and the Scientists" by John S. Sargent, provides a glimpse into the Koreshan reality in much the way science fiction creates fictionalized theories of reality, but *this* reality was believed. Halley's comet, Teddy Roosevelt, astronomy, ancient knowledge mixed together in an odd stew of thought—the principles of physics, astrophysics, astronomy are all turned inside out in this almost Alice in Wonderland cosmology.

THE FLAMING SWORD
Halley's Comet and the Scientists

While our first citizen, "our Teddy," has been dashing meteor like through the palatial realms of the Old World, astounding and charming alike both king and proletariat, the New World has been witnessing and wondering at the doings of a strange visitor in the starry heavens—Halley's comet. To intimate that there is the remotest relation between these two events, would be to plunge the subject into the wildest kind of unprovable speculations. Yet they could hardly transcend the exaggerated descriptions and predictions of the "scientists," and the unwarranted fears of the superstitious concerning this strange wanderer of the gem-bedecked skies of the earth.

What is it? and whence is it? are questions that, like the scientists' own theory of the wasted energy of the sun, float off on the endless ether from which no echo even deigns reply. Comets come and go, and our learned men—despite their boasted erudition and superior instruments of observation—know little more of them than did the ancient Chaldeans, further than to have established the periodicity of some at least of the more noted. Halley's comet is one of these; it is found to return to our view every seventy-five years, and, by the magic of the magnificent distances figured into the Copernican system, it is accredited with darting through space at the impossible rate of forty-six miles a second. This rapid progression necessitates accrediting it with an orbit of billions of miles in circuit, and the alleging its tail to be streaming out for 20,000,000 miles.

These scientists laugh at the credulity of those who think everything possible with God; but the credulity that can accept such impossible distances and dimensions in the name of "science," is equally absurd. They have frightened nervous people with the suggestion that the comet might run afoul of the earth, or suffocate us with poisonous gas of its tail—as it swished by us in its transit across the face of the sun. But is has passed and nothing serious has occurred; even the closest scrutiny with telescopes has failed to disclose anything interfering with the light of the sun, other than a faint bar across its disk, and a slight appearance of the sunlight as that just preceding an eclipse. Nothing more; no solid body that could shatter the earth or plug a hole through its shell, had it struck us. Add to this the fact that stars have been seen through the comet's nucleus, and it will show that its body is about the consistency of the aurora borealis or northern lights. All of these facts go far to prove the assertion of the Koreshan Science of Cosmogony,—that comets are simply aggregations of energy on their way from the earth to the sun.

Now Sargent analyzes the etymology of the word "comet" arguing that the word is to be understood *literally* instead of *figuratively*.

> If the learned men of our day had not been so wise in their own conceit as to discredit the knowledge of the ancients, who gave names to the orbs of heaven, they might have very sensibly inquired why these sky wanderers were given the name of comets. Perhaps they did; but if so, they credited ancient knowledge with the same superficiality they possess. As the word comet means hair, they naturally concluded that the name was given because the long tail streaming out behind was suggestive of the unshorn locks of a Nazarite. But whoever it was gave names to the things of the heavens and the earth, had a deeper insight into the functions of these various things than that, *knowing that the universe was an unincubated man, with an integral relationship of its various parts to the corresponding parts of the incubated man—the human form.* (Italics mine.) So the name of comet was given to these flying aggregations of energy, from the Latin word cometas—meaning hair, for the very good reason that in function, their substance is to the physical universe, what hair is to the human body....

The author returns to a Koreshan explanation of the energies which project comets from the earth toward the sun's focalization.

> The sun throwing out its various energies, as photoic, caloric, scotoic, and cruosic, etc., in spirals, necessarily such an ascending body finds its way along the demarkation lines between these outgoing energies. This makes the comets describe what appear to be orbits, and passing out of sight into the sphere of hydrogen, our "scientists" suppose them to have gone on a long journey in stellar space. If the comet went to the moon, she, being feminine, could with a stick-pin don it as an ornamental addition to her coiffure, with which to excite our admiration. But the sun being male, he simply burns it up to replenish his energies, and Mother Earth busies herself accumulating another supply, to be clipped seventy-five years hence, to take the same course through the heavens. In the meantime there will be many other clippings, which, being somewhat different, will so appear, and take other routes through the heavens.

Such eloquent discourse in the service of an impalpable dream, a vision of a universe that is not! The view of *Flaming Sword's* editors, writers and readers stands as a testament to theory based upon "revelation" and "reasoning" rather than upon concrete reality apparent and provable in the physical universe.

From Mary Proctor through Flammarion to Dr. See or "Koresh", from orthodox, legitimate astronomy through fear-mongering astronomy to the pseudo-
religious and pseudo-scientific. Halley's comet's 1910 apparition mirrored the
wisdom *and* folly of human thought, a richness of world views, a richness of
voices delightful to hear of today.

III
Four Great Figures

13. "THOU ROOSEVELT OF THE HEAVENS"

Because Halley's comet was visible for so long —from April to June, 1910—advertisers, cartoonists, poets, composers, postcard manufacturers, editorialists, and the like had the opportunity to portray and discuss the comet from every point of view and use it in every conceivable way, from the plausible to the far-fetched.

What had Theodore Roosevelt done to make association with Halley's comet apt, especially since he had left the presidency over a year before the comet brushed by the earth? Nineteen-ten was a "blazing" year for Halley's comet and for Roosevelt. The connection between the two lights of the world was manufactured in political cartoons, humorous poems, and editorials portraying the comet and the man as one. Having declined the Republican party's nomination in 1908 after nearly eight years in office, Roosevelt selected Secretary of War, William Howard Taft, as his successor. Taft easily won the election and on March 5, 1909, President Roosevelt stepped down. Less than three months later he left Oyster Bay, New York, to begin a long-awaited African safari.

A noted naturalist, conservationist, historian, and hunter, Roosevelt was the ideal person to lead the eleven-month safari collecting big and small mammal specimens for the Smithsonian Institution. By mid-May of 1909 he had bagged six lions, two giraffes, a hippo, zebra, many varieties of antelope, and a rhinoceros. The expedition's 260 porters, guides, and the most proficient American naturalists and taxidermists of the day trekked southward. At Lake Naivasha, Roosevelt hunted hippopotamuses from a rowboat. He shot elephants on the slopes of the Nile, then entered the Belgian Congo, where the President shot nine rare white rhinos. By December the group had collected more than 500 large mammals— "The most noteworthy collection of big animals that has ever come out of Africa," Roosevelt wrote—and a total of over 11,000 animal specimens.

In addition to overseeing the activities of the entire expedition, Roosevelt met with local government officials and residents, reviewed several books for American magazines, kept up his massive and eloquent correspondence, and read voluminously as he always had. Throughout the trip Roosevelt's interest in history, military affairs, feats of engineering, and colonial enterprise was satisfied by penetrating discussion with knowledgeable locals. He kept America informed of his adventures in Africa through his own written accounts which appeared regularly in *Scribner's Magazine.* By the time he returned to America the reports had been collected and published as the very popular *African Game Trails: An Account of the African Wanderings of An American Hunter-Naturalist.*

Having slowly progressed down the White Nile to Khartoum in what was then called the Anglo-Egyptian Sudan, the expedition was concluded in March of 1910. Roosevelt emerged from the jungle and the delights of big game hunting and scientific inquiry to find that though he had been absent from the politics and public scene for more than a year, interest in him had *grown* rather than diminished.

A few short days after his arrival in Cairo, Egypt, Roosevelt delivered an inflammatory address at Cairo University on the subject of the Nationalist movement in Egypt to replace British rule with native rule. The sore point of his address dealt with the unpunished assassination during the previous month of Premier Butros Pasha, hated by the Nationalists for supporting the British administration. Of the incident Charles Morris wrote, "...the English, for some reason hesitated to act promptly. There was no hesitation in Theodore Roosevelt. Murder was murder, and could not be glossed over. Though there were forebodings of trouble if he should refer to this event, they failed to affect him." The speech was studded with remarks such as these: "The type of man that turns assassin is the type possessing all the qualities which alienate him from good citizenship, the type producing poor soldiers in time of war and worse citizens in time of peace." Emerging from the African hunt, Roosevelt reentered the world political arena, his voice as straightforward, fearless, and uncompromising as ever.

Roosevelt had planned to return directly to America when the safari was over. An invitation to deliver the prestitious Romanes Lecture to the scholars of Oxford changed his mind. Proceeding through Italy, Roosevelt was moved to decline a visit to the Pope because of unacceptable conditions imposed by the host. This set off a second political bombshell, placing the hunter once again in the center of controversy. He next traveled to meet the Kaiser in Germany, then stopped in the Netherlands, then to England to deliver the Romanes Lecture at Oxford University and, his visit coinciding with King Edward VII's death, represented the United States in the ceremonies. Characteristically, his remarks on royalty were to the point. "...the King's lack of political power, and his exalted social position, alike cut him off from all real comradeship with the men who really do the things that count..." And, in a lecture at the Guild Hall in London, he instructed the leadership of Europe on the principles of governance for the vast colonial empires they had amassed.

Time after time political cartoonists, "poets," and editorialists made the connection between the light of the world blazing his way across the day's press and the celestial visitor. By 1910, the United States had begun its ascendancy among the powerful nations of the world, and Roosevelt's speeches, actions, and admonitions left his European hosts shaken and fearful. An American editorialist noted that Americans themselves need have no fear of their former President:

> Popularity that approaches the bounds of hysteria is a dangerous blessing in America. But that is no ground for mistaking the sincerity with which the nation welcomes the return of its former President. The vital question is, what Mr. Roosevelt plans to do with his popularity. There are those who look with apprehension on the return of Mr. Roosevelt from his year of seclusion among lions, crowned heads and newspaper correspondents. Manuals of universal history have been searched to reenforce the memory of alarmists with parallels in the lives of Julius Caesar, Napoleon Bonaparte and any other man who, leaving his country a republic, returned to make it an empire.
> *True, we have not always taken these alarmists very seriously, but now that we have escaped the comet—or, the comet has escaped us—we shall be subjected to new editions of direful prognostications.* (Italics mine.)

The booming optimism, so amply expressed by President Roosevelt himself, provided the tone for the remainder of this editorial that tied Halley's comet to Roosevelt. The editorialist remarks, for example:

> *The popularity of former President Roosevelt—we like that title better than the belligerent 'Colonel'—is certainly unparalleled. Yet, study the national mind and you will find no cause for alarm. This extravagant popularity does not threaten our republican institutions. It is gratuitous and worse to argue that because we honor a man who stands for democracy, democracy itself is threatened. The American people is not plotting against the constitution because it welcomes Mr. Roosevelt.*

The comet is now visible at Washington.

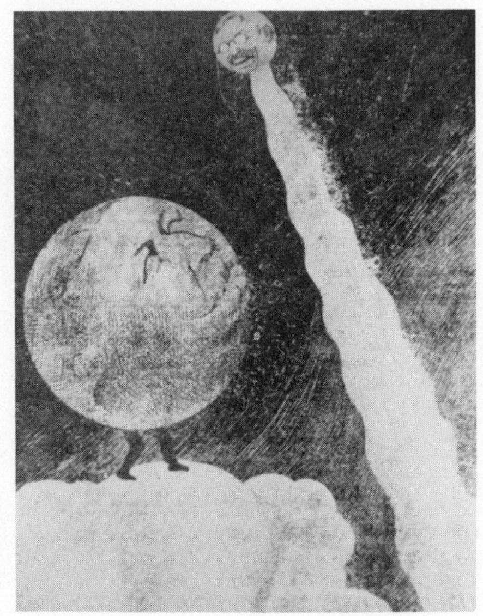

46

Nonetheless, the fear that the editorialist described and dismissed was widely expressed. Three cartoons told similar stories of the apprehension with which Roosevelt's return to America was regarded, two seen *from the point of view of President Taft.* It was widely thought that Mr. Roosevelt might well be returning to reclaim the presidency, even though he had renounced a third term. The *Review of Reviews'* "Cartoon of the Month" captioned "President Taft and the Comet" believes that Mr. Taft regarded or should have regarded Mr. Roosevelt with much the same apprehensions as those with which the world regarded Halley's comet. "The Comet is now visible in Washington" said McDougall in *The Philadelphia Telegraph* and *The Literary Digest.* Taft, clothed in his pajamas, looks out into the night and sees the shining teeth and glasses descending upon him in the figure of Halley's comet.

Even in the august ceremonies preceding the ex-President's Romanes Lecture, Oxford's Chancellor addressed the Vice-Chancellor in Latin verse as Mr. Roosevelt took his place to be presented:

> Behold, Vice-Chancellor, the promised wight,
> Before whose coming *comets turned to flight,*
> And all the startled mouths of sevenfold Nile took fright.

Roosevelt-Halley's comet poems appeared in many magazines and newspapers. A "Salutation to Halley's Comet" begins:

> Of gases all compounded,
>   Forever chasing far,
> And yet engaged in doing
>   Naught in particu-lar;
> The solar system putting
>   At sixes and at sevens—
> We gladly, proudly greet thee,
>   Thou Roosevelt of the heavens!

A parody of "Twinkle, Twinkle, Little Star" appeared in the *New York Tribune* for May 15th, entitled "Roosevelt Praises the Comet:"

> Twinkle, twinkle, Halley's C.,
> How your action pleases me;
> Marathoning past the sun
> Like a diamond from a gun.
>
> Scooting, sparkling through the void,
> You affright the asteroid;
> Foreign worlds, despite their fears,
> Herald your approach with cheers.
>
> Brave Rough Rider of the sky,
> Fain would I cut loose on high,
> Blazing astral paths like thee
> (Fixed stars don't appeal to me).
>
> Comet, dwelt I in the sky,
> Much like you I'd pack and fly
> Where the Heavenly Game Trails are—
> I would be a Shooting Star.
>
> *(author unknown)*

Additional poems complete this unexpected anthology of Roosevelt and Halley's comet verse:

The New Comet

Tearing through dark jungle lands, abeaming with delight;
Hunting nature fakers down, and putting beasts to flight;
Entering the zone of death, triumphant in his gait;
Outpointing Wanderobo guides, tied to the tail of fate;
Deploying on the mystic Nile, waking up the Sphinx;
Obliging with some burning brands on Egypt's social kinks;
Rumbling o'er the inner sea, just one step nearer home;
Entertained and lionized and passed along to Rome.
Rousing all the cable men to flash wherefore and why
Of Vatican refusal, of question and reply;
Of religious controversy, advice and persiflage,
Sweeping into column rules the axioms of the age;
Enlivening a breathless world, waiting for the next
Veni, vidi, vici of the modern god of text;
Emperors and kings fade out before his brilliant ray
Like another comet he is flaring on his way;
The worst is yet to hit us—some time in June, they say.

Percy Shaw
New York American
April 6, 1910

Roosevelt was to return to America in June. His country gave him a tumultuous welcome.

And, finally one, the other, and both Halley's comet and Theodore Roosevelt.

Fear Not!

His head is broader than the sphere
    On which we mortals dwell,
And through the shuddering atmosphere
    He rushes on pell-mell;
The planets shriek, the mountains melt,
    His path, the sun shrinks from it,
Do we refer to T. Roosevelt?
    No, no; to Halley's comet!

A gaseous vapor marks his track,
    Woe, war and trouble follow;
Who sees him cries, "Alas! Alack!
    He has us frightened hollow."
He's lighter than a sparrow's beak,
    Though bigger far than Jumbo,
Of Halley's comet do we speak?
    No, no; of Bwana Tumbo!

But don't be scared, O timid folk;
   Let courage not desert you,
Though big and fierce, he's but a joke;
   He has no power to hurt you.
Smile as his shape across our brow
   He sweeps in his mad waltz, oh!
We speak of Halley's comet now,
   And of T. Roosevelt, also!

Paul West
New York World
Apr. 19, 1910

Following the years of Roosevelt's ascendency and in the days of brightest glory, Halley's comet served well as a celestial image to mirror his earth-bound achievements.

Proposing the health of the ex-President, the Mayor of Rome called him, "…one whose character and work had an effect upon the civil progress of humanity. Men of his calibre are beyond the limits of country. They belong by right to civilization." Like Halley's comet in 1910, Roosevelt appeared to be beyond the limits of country, belonging to civilization.

14. MARK TWAIN (November 30, 1836–April 21, 1910) AND HALLEY'S COMET AT PERIHELION (November 16, 1835–April 20, 1910)

In the public's opinion Halley's comet and Theodore Roosevelt were emblems for one another. Yet, at the time of his death on April 21, 1910, Mark Twain's life span coinciding with the long "year" of Halley's comet was mentioned only in a letter to the editor:

MARK TWAIN AND HALLEY'S COMET

To the Editor of *The New York Times:*
   I wish to draw your attention to a peculiar coincidence.
   Mark Twain, born November 30, 1835.
   Last perihelion of Halley's comet, November 16, 1835.
   Mark Twain died April 21, 1910.
   Perihelion of Halley's comet, April 20, 1910.
   It so appears that the lifetime of the great humorist was nearly identical (the difference being exactly fifteen days) with the last long "year" of the great comet.

R. Friderici
Westchester, N.Y., April 22, 1910

Everyone knows Mark Twain was born with the comet and died when it came again. By a strange twist of fate, it is also the sole fact widely known about Halley's comet! The true legend of Twain and the comet took longer to be noticed.

In *Sam Clemens of Hannibal: The Formative Years of America's Great Indigenous Writer,* Dixon Wecter wrote:

> To Mark Twain, schooled in the omens and clairvoyance of the backwoods, this apparition seemed mystically bound up with his own span from birth to death. "I came in with the Comet," he said, "and I shall go out with the Comet"—this he did. In an unpublished version of *The Mysterious Stranger,* set not in medieval Austria but in Hannibal, he has the Stranger ask: "How do you know, when a comet has swum into your system? Merely by your eye or your telescope—but I, I hear a brilliant far stream of sound come winding through the firmament of majestic sounds and I know the splendid stranger is there without looking."
>
> And as if unconsciously to fulfill the first date of his destiny, he came forth untimely from his mother's womb, a seven-month child, on November 30, 1835.

Then Wecter tells this wonderful little tale:

> The infant Samuel barely survived that bleak winter and the next year or two. According to Jane, "When I first saw him I could see no promise in him. But I felt it my duty to do the best I could. To raise him if I could. A lady came in one day and looked at him she turned to me and said you don't expect to raise that babe do you. I said I would try. But he was a poor looking object to raise." Many years later, when his mother was in her eighties, they had a conversation about those times, in which Sam asked, "I suppose that during all that time you were uneasy about me?" "Yes, the whole time." And then, deliberately offering a gambit such as Jane Clemens loved, "Afraid I wouldn't live?" A long pause for reflection—"No, afraid you would."

On April 20, 1910, just one day after Halley's comet had passed perihelion, the point closest to the sun,

> Mark Twain's grey, aquiline features were molded in the inertia of death for long hours, while his pulse sank lower and lower. But late at night he passed from stupor into the first natural sleep he had known since he returned from Bermuda. On the morning of April 21 he woke refreshed, even faintly cheered and in full possession of all his faculties. He recognized his daughter Clara, spoke a word or two, and feeling himself unequal to conversation, wrote out in pencil, "Give me my glasses." They were his last words. Laying the glasses aside, he sank first into reverie and later into final unconsciousness at 3:00 in the afternoon. At 6:30 that evening Samuel Langhorn Clemens, "Mark Twain," died painlessly of angina pectoris at Redding, Connecticut.

Fort Meyers Press,
April 28, 1910

Twain's death inspired heart-felt notices. Widely and warmly recalled, his death was marked by world-wide tributes. Every social class paid tribute to Twain's great contribution to the world's wisdom and humor, expressed sadness at his passing, and demonstrated the deep regard in which the beloved author was held. Beneath a black-bordered photograph, *The Review of Reviews* summarized his career and impact.

Born November 30, 1835   Died April 21, 1910

Mark Twain (Samuel L. Clemens)

If America has produced greater men of letters than Mark Twain, certainly no other writer has held for such a long time so much of the esteem and affection of contemporary Americans. For fifty years Mark Twain has been making the world glad. Mr. Clemens' life story is as picturesque as the quality of his humor. Born in 1835 at Hannibal, Missouri, he was apprenticed to a printer at twelve years of age; in early manhood he was pilot of a Mississippi steamboat; at twenty-seven he was editor of a paper in a Western mining camp, and then a real miner himself. In the decade following 1870 he became famous as a humorist. "The Jumping Frog," which more than any other single story began his fame, appeared in 1867. Forty-one years later he was still hard at work writing his autobiography. Two or three years ago Mr. Clemens set up his household goods at Redding Ridge, Connecticut, where he built a stately house in a lovely countryside.

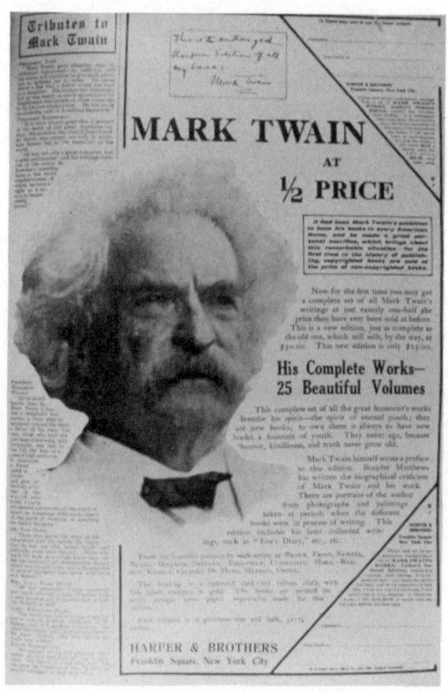

Harper & Brothers, publishers of a twenty-five volume *National Author's Edition* of Twain's works, placed an advertisement in *Harper's Weekly* of May 21, 1910, with a column alongside the ad entitled "Tributes to Mark Twain." Tributes from Presidents Taft, Roosevelt, and Wilson, a professor, and three notable writers expressed high regard for Twain and a deep sense of loss at his passing. Here are remarks of two mourners:

> *Theodore Roosevelt:*
> It is with sincere grief that I learned of the death of this great American author. His position, like that of Joel Chandler Harris, was unique, not only in American letters, but in the literature of the world.
> He was not only a great humorist, but a great philosopher, and his writings form one of the assets in America's contributions to the world of achievement, of which we have a right as a nation to be genuinely proud.

> *Mrs. Julia Ward Howe* (Author of *Uncle Tom's Cabin*):
> He was personally highly esteemed and much beloved, a man of letters with a very genuine gift of humor and of serious thought as well.

The truth is, Twain *did* come in with the comet and go out with it. And one of his fictional characters in an unpublished version of one of his most famous stories said that he could hear the sound of comets. Beyond that, there is just a pleasing appropriateness in having Twain grow tall in our imaginations, tied to Halley's comet.

Halley's comet's apparition is one of those rare events which causes the world to view itself against the night sky and the solar system. And in the foreground of this most dramatic of the solar system's displays walked America's most flamboyant president and its most beloved writer. The comet suggests a universal dimension in their personal lives and demonstrates the power of emblems, symbols, and images to create powerful associations in the mind, in the national consciousness.

15. THE KING IS DEAD: LONG LIVE THE KING: HALLEY'S COMET, KING EDWARD VII, AND KING GEORGE V

Comets have always been popularly believed to be omens of impending change, especially the deaths of kings and rulers. By 1910 such superstitions were largely part of the past, but the rather sudden death of King Edward VII on May 6, 1910 brought the belief back to mind.

Edward VII, King of Great Britain and Ireland, and Emperor of India, died at Buckingham Palace on the night of May 6, 1910 at the age of 69. The cause of death was acute bronchitis and pneumonia. His illness lasted only five days. On April 26, King Edward had returned to England from a vacation on the continent in good health except for throat trouble which had afflicted him for years.

Shortly after his accession to the throne nine years before, the King had consulted eminent specialists about a throat cancer, and an operation had been performed at the time. During his vacation he had again consulted physicians and was treated again for the old trouble. Dr. St. Clair Thomson said another

operation was needed. Later it was discovered that the lungs were more seriously affected than the throat; the operation was postponed.

At week's end, King Edward went to Sandringham to inspect improvements
being made there. The weather was cold and damp, and he caught a severe
cold. Returning to Buckingham Palace Monday, his condition became serious.
Pneumonia developed soon after. Despite constant care, the royal patient's
condition rapidly worsened. Alarmed by the King's condition, the physicians
called in Dr. St. Clair Thomson and Dr. Bertrand Thomson, eminent throat specialists. Their efforts were fruitless, however, and by Friday morning, May 6, the King
was on the verge of death. By the close of day oxygen was administered, only
serving to prolong life for a few hours.

During the day the servants of the royal household had changed their
scarlet liveries for black costumes, and an urgent summons was sent for the
royal family. All day the Queen, other members of the royal family, and the four
physicians remained in the King's apartments. Up until a few hours before his
death the royal patient was conscious and, while aware that the end was near,
he declined to go to bed. While sitting in his chair King Edward was seized with
frequent violent attacks of coughing. Twice His Majesty fainted. Even in those
last hours he transacted some public business. By evening the constant strain of
coughing and difficult breathing so affected his heart that the left ventricle was
failing. Oxygen was giving no relief. By six o'clock his condition was critical. In the
evening the King asked if his race horse, Witch of the Air, had won at Kempton
Park. Told that he had, the King smiled.

The Archbishop of Canterbury visited the Queen in the morning and called
before dinner. Attending a large charitable meeting in the evening the Archbishop
said, "Sickness and sorrow and the great issues of life and death level all earthly
barriers. I ask you as Christians to stand for a minute in silence and lift up your
prayers for our sovereign hanging this moment between life and death." The
audience rose and, after a minute of silence, joined in the Lord's Prayer. The
Archbishop returned to the palace at 9:30 in the evening.

Finally, the King became so exhausted that he retired, uttering his last
words which immediately became famous: "Well, it is all over, but I think I have
done my duty." Then he became unconscious until about 10 o'clock, rallying just
long enough to recognize family members, but still unable to speak. After a few
moments King Edward VII sank into unconsciousness again and remained coma-
tose until he died at 11:45 p.m. Five minutes later the four physicians in attend-
ance issued this announcement: "May 6, 1910, 11:50 p.m. - His Majesty the King
breathed his last at 11:45 tonight in the presence of Her Majesty Queen Alexan-
dra, the Prince and Princess of Wales, Princess Royal, the Duchess of Fife,
Princess Victoria, and Princess Louise, the Duchess of Argyle. (Signed) Laking,
Reid, Powell, Lawson."

Sir Francis Knollys, private secretary to His Majesty, officially announced
the King's death half an hour after King Edward had passed away. Sir Francis
walked into a room where the newspaper reporters were waiting and said,
"Gentlemen, His Majesty is dead."

The first official act of the new King, George V, was to inform the Lord
Mayor of London, in accordance with usage, of the death of the King. The
informal message read, "I am deeply grieved to inform you that my beloved

father, the King, passed away peacefully at 11:45 tonight. (Signed) George."

People in Bermuda, Jamaica, the Bahamas, and other British colonies in the Windward Islands as far south as Trinidad, British Honduras, and British Guiana observed the long-expected Halley's comet suddenly become visible the same night of King Edward's death. That night the fort at Hamilton, Bermuda, began to fire a salute of 101 guns at 12:30 a.m. in honor of the new King. Every two minutes a discharge was fired. The comet became visible at 2 o'clock in the morning. A red tinge was noted in its tail. The last salute was fired at 3:52 a.m. exactly. As the guns died away, observers at the fort saw a sudden flare at the end of the comet's tail. The head also glowed like a ball of red fire. The glow lasted for five minutes and was seen by Negroes working on the docks. Overcome with terror, they fell to their knees and prayed; some thought the end of the world was coming. It was impossible to get the stevedores to go on with the loading of the steamship, *Bermudian.* They believed that the glow and flares showed that war would break out during King George's reign and that a great calamity would befall the earth. Some fell speechless with fear. Some worked themselves up in the paroxysms of religious zeal to a frenzy. Until the comet faded from view and the daylight of Saturday had broken, they could not be induced to return to their work.

Later, notice of the coincidence of the King's death and the comet's presence appeared in several places. An obscure English magazine whose topic was events in the vast English empire, *East and West,* had these comments on the matter:

> In a scientific age the appearance of a comet, which approaches the earth only once in seventy-five years or so is worthy of record among current events. Perhaps it caused more stir among the people in bygone times, when it was believed to bring disaster in its train, than it does in our enlightened age. India, however, is not altogether enlightened, notwithstanding the efforts of so many newspapers to diffuse knowledge and refinement throughout the land. The educated leaders of society have not felt ruffled in their minds by the appearance of Halley's comet last month, and therefore the masses too have looked upon it with more interest than consternation. *In the British Empire those who are inclined to associate terrestrial events with celestial phenomena may record the curious coincidence of the death of King Edward with the appearance of Halley's comet. But of course not many will openly profess to believe that kings and queens die only when something strange occurs in the heavens.* (Italics mine.)

Saturday, May 7 was the day set by the City of Liverpool for its May Day Procession. In a poem written on the occasion, J. Lewis Milligan penned these two stanzas to open his poem honoring the dead sovereign:

MAY THE SEVENTH, 1910
(Liverpool's "May Day")

Last night the west was weird; stupendous clouds,
Like a long range of lunar mountains, stood

Against a tragic sky of ghostly grey;
While overhead hung airy continents
Of black portentous clouds, from whence there came
Bright, stinging darts of icy rain. The air
Seemed all posses'd with howling wraiths, as I
Headed my way, half-blinded, toward the town.
I saw the city lights like fallen stars
And men, like ants, creep through its glimmering ways;
And, as I joined them, to my ears there came
The solemn tidings of a stricken King.
Throughout the night, as on my bed I lay,
Fitful I slept; the furies of the north
Went wailing through the streets, as they did sing
A canticle of hell! At last I rose
And drew aside my casement blind, and lo!
Adown the west there blazed the morning star!
Fair Venus! Daughter of the Sun! Methought,
Had I ne'er lived for anything but this,
Through all these stormy years—'twere not in vain!
When on my raptured sight the comet gleamed,
That fiery Absalom, which boded ill
To Emperors of old—and from my heart
I breathed a fearful prayer for Britain's King. (Italics mine.)

Surely, from the days of early Rome and the Caesars, and most likely from the time even before writing, comets have been taken as omens presaging the deaths of kings and princes. The assassinations of Caesar, President Garfield and Czar Alexander II, the death of Napoleon, and numerous other heads of nations were believed to be prophetically announced by comets. Even Halley's comet itself is closely tied with England's history. The justly famous Bayeux tapestry, the work of Queen Matilde (or commanded by her), wife of the Norman William the Conqueror, shows King Harold of England cowering in fear on his throne while Halley's comet hangs over his castle. Naturally, the tragic death of King Harold and the conquest of England by the Normans were believed to be foretold by the comet.

Eight hundred forty-four years later, in an enlightened age, another king died during the comet's reign. The measure of the change in outlook is reflected in the absence of outbreaks of hysteria. The few references to the coincidence were subdued. Yet, coincidence or portent, the king of one of the world's greatest empires died with Halley's comet in sight, and the coincidence is worthy of mention, shedding light on one of the great interfaces between the comet and humanity.

16. HALLEY'S COMET AND POPE PIUS X

On May 28, Pope Pius X visited the observatory in the Vatican Gardens to see Halley's comet. He and his retinue were received by the observatory director, Father J. G. Hagen. A *New York Times* correspondent accompanied the party. The Pope was interested in the view of the comet, but thought it did not justify all the commotion in the old and new continents.

His Holiness said that one night a pharmacy in Rome received an urgent call for oxygen to keep a dying man alive. Not one druggist in Rome had any bottled oxygen left. It had all been bought by people who were frightened that the gases from the comet would kill them. They planned to use the oxygen to keep themselves alive until the earth had passed through the comet's tail. And while no ill befell any of those who bought oxygen, their ignorance was the instrument of another's death.

IV
The Comet in Print

17. HALLEY'S COMET IN CARTOONS AND HUMOR

Cartoons of every sort—political, social, sports—used Halley's comet to depict the foibles of mankind and the day, and to convey light-hearted insights. The cartoons which portray the comet were simple, naive, and unsophisticated. A great deal of the comic material was predictable—the comet and romance, the comet as an excuse for coming home late, the reaction of inebriates, the comet and fear, the end of the world—but there was much that was truly novel and inventive. The arrival on the scene of the automobile and the even more recent appearance of the "aeroplane" called forth several cartoons pointing out the incredible speeds now possible on earth. Why, a car could travel up to twenty-five miles an hour and craft heavier than air could cross the English Channel!

THE GREAT AMATEUR
Aviator: "Marvelous flier! and he does it for love!"

The June-moon-night-is-for-romance theme appeared in many cartoons depicting young lovers sitting out at night, blissfully absorbed in each other's company and ignoring the very object they sat out to observe.

The Halley-Luia comet. Oh Horace! Aren't the comets lovely tonight?   *Waiting for the end of the world.*

"Lame excuse" cartoons and humorous pieces latched onto the comet for their laughs. In an installment of William F. Kirk's popular "Little Bobbie's Pa," Pa used the comet's passing as an excuse to spend the night of Friday the thirteenth with his "gang." In exchange for granting him the privilege, Ma gets $20.00 to buy herself a new summer hat.

Ware are you going? sed Ma to Pa last nite wen he started for to go out.

Well, sed Pa, I will tell you ware I am going; I am going out to get a peek at the comet. Reemember, Pa sed, this is Friday, the 13th, & sumthing seems to tell me that it is a fateful nite. Sumthing tells me, Pa toald Ma, that the comet is going to raise the dickens tonite, & I want to be right on the ground when trubbel cums....

Jest a minnit, Ma sed, beefoar you go. You see, light of my life, I doant want you to fergit that littel matter of $20 wich you was going to leeve in the sugar bowl. You understand, dear, I doant want to do anything wich will interrup you in yure searches after astronomy truths, but you perhaps recall about that littel $20 for my new Summer hat. Knowing that you have the $20 now, Ma sed, & knowing as I do that after you have been out with yure gang you are likely not to have the twenty, I doant think it wud be a half bad idee for you to unload right now.

So Pa gaiv Ma the $20, and then he went out to look at the comet. I wanted him to take me with him, but Pa wuddent.

The next morning wen Ma & me was eeting brekfast, Pa caim hoam. He must have been looking rite at the comet a long

time, beekaus his eyes was the saim color, red, & he walked like
he had a stiff neck, too.

Well, sed Ma, give account of yureself. Gladly, gladly,
daughter of the sun, sed Pa. It was the finest comet wich I have
evver saw. It was gorgus.

Describe it, sed Ma.

And Pa philosophizes:

As I was saying, sed Pa, we stood & looked at it by the
hour, & as we stood & looked at it, the greatest feeling of awe in
the whole world caim steeling oaver us. It was then, sed Pa, that
we reelized the poor little earth worms we are. Think of a wun-
derful body, milyuns & milyuns of miles in length, coming near to
the erth & then steeling off aggenn into space, the boundless
space of the grate universe, the il-limitabel area of the All. That
is what the thoughts was that was chasing through our brain,
sed Pa, wen we saw this grate comet.

Pa goes on with such thrillingly described emotions. Ma is not taken in and
after Pa finally coughs up a doubtful description, Ma sed, "Well, go to bed and
stay thare." Another sure topic for a cartoon was the confused drunk gentleman
asking a policeman while pointing at a street light, "Orfesher, 'blige me! Whish
one er thosh ish Halleysh comet?" (from *Puck*)

Stayed up to see it.

Political Humor

Several political cartoons using Halley's comet appeared from March
through June of 1910. Most compared the political luminary of the moment to the
comet. The few that went beyond this simple association are of considerable

interest. By far the figure appearing most frequently with the comet was Teddy Roosevelt. The story of that association has been told in an earlier chapter.

One other political figure of great international importance, Kaiser William II of Germany, appeared in two cartoons, both making the same point.

HALLEY'S COMET.
William II—"The end of the world? Impossible! I have given no such order."

Published in Pasquino *(Turin, Italy) and reprinted in* The American Review of Reviews.

The cartoonist's portrayal of William—erect, firm, stern, sword prominently visible—reinforces the caption comment. It is taken for granted that the Kaiser has the power to destroy the world and that upon his order it would be done.

WILLIAM THE CONQUEROR
(An English Dream)

The cartoon depicting William the Conqueror appeared in the Literary Digest, *alongside an essay entitled "German Unpopularity Analyzed."*

"Halley's comet appeared in 1066—when William the Conqueror took England. Halley's comet is here today."

The theme of war with Germany was in the air. The 1066 apparition was often mentioned in the press, as indeed were many other past appearances of the comet. The specific theme of the conquest of England and current affairs appeared in just these few items.

A note taking the same view of the comet's place in English history appeared in the *Observatory:*

> A correspondent kindly sent me the following suggestion a few weeks ago. It is not his own, but he says that it is only fair that the authorship should remain anonymous:—
>
> Halley's comet, that which will be visible in 1910, is the one that was seen in 1066, and it will be observed in precisely the same position now as then, *i.e.*, so I am told. In the first case it was supposed to be an evil omen foreshadowing the defeat of the English—then Saxons; and this was correct, for the Saxon Army under Harold was defeated at Senlac Hill by William the Conqueror and his Normans.
>
> Now in 1910 a German invasion is predicted. Will it be with the same result for the English? And the Germans are making such strides in aerial navigation too.
>
> Remember in *both* cases the date contains a "10"—1066 and 1910.
>
> My correspondent adds an original idea of his own, viz., that the name of both aggressors, actual and hypothetical, is *William.*

Combining numerology, divination by names, and astronomy with current history, the correspondent raises the same questions the cartoonist asked. Will Germany engage England and the world in a war? Is history to tragically repeat itself? The editor's closing remark struck an ominous note. In Parliament England's preparedness for what was taken to be signs of a coming war were being debated.

> Since I received his letter we have had the Naval debate in the House of Commons, and perhaps the jest is a little too near earnest to retain its full flavor.

Halley's comet had appropriate associations for the nation and person threatening to use its might and power in a war. Portent, or just common sense? Four years later the political debates and diplomatic accusations which flew from nation to nation throughout Europe came to a sad and tragic climax in the declarations of war and the outbreak of hostilities, the opening of the Great War foreseen in these cartoons of Halley's comet.

Astronomy Humor:
Or, "If the Human Frame Has Its
Alimentary Canal, Why Not Mars?"

A brand of humor which portrays the astronomer as an unrealistic goof ball, a head-in-the-clouds sort of fellow, accompanied Halley's comet. Wex Jones' caricature of astronomers in "Our Hurried Caller" employs many of the tricks of

this sort of joking. In a derisive tone he picks on the "prophesy" of astronomers—
including even Halley, whose prediction was accurate!

> [Halley] merely made a prophecy that the comet would
> return in about seventy-five years. He would be safely dead and
> beyond the reach of scoffers. In the meantime he was sure of a
> great reputation while he lived. This is on the plan adopted by
> those who prophesy the end of the world. They put the event far
> enough ahead to enjoy the little notoriety before they are shown
> to be mere guessers, and bad ones at that.

Now this is mean-spirited humor! By presenting parodies of actual astronomers'
conflicting scientific opinions, Jones raises the question, if science is the way to
the truth, then why is there so much disagreement among the word's eminent
astronomers? The fact is, this attitude on the part of many journalists toward
scientific theory is with us today; witness what the news reporters say to the
meteorologists on radio and television. All those stale jokes treating the weather-
man as if he made the weather instead of just forecast it. A few of Wex Jones'
bad jokes go a long way:

> PROFESSOR LOWELL—The comet will not hurt anyone.
> Its disposition is naturally affectionate, and if children are warned
> not to pull its tail it will pass us by without scratching. The comet
> clearly proves that there are canals on Mars and that the annual
> tonnage on the Grand Trunk Canal must exceed that carried on
> the Grand Canal at Venice. If the human frame has its alimen-
> tary canal, why not Mars?
> PROFESSOR JUGGINS of Greenwich Observatory—
> There is absolutely nothing to fear from the tail of the comet. It is
> like a motor car. If the car hits you, you are a dead one, but you
> can cross the tail of gasoline vapor with impunity.
> DR. LUDWIG SPIEGELBROTZEN of Berlin—The comet
> has no tail. It is merely a beam of light that we see. If I am wrong
> in this surmise there will be no one left to say, "I told you so."
> The end will be sudden and painless.

Three other astronomers appear on the scene to comment foolishly about Halley's comet. A cartoon displaying a lion, an elephant, a cow, and a horse, all
laughing, accompanies the story. Presumably the message is that the words of
the astronomers sound funny to even the beasts. Wex Jones got a lot of mileage
out of Halley's comet. He also wrote a piece entitled "The Diary of a Comet."

Ethnic Humor

In 1910 it was still common to find ethnic humor in newspapers and magazines. Foreigners and blacks were always good for a laugh, deplorable as that is.
The Swedish servant, the wise old darkie, and the Pullman porter were among
the figures whom people chuckled at when Halley's comet flew by.

Frank L. Stanton created an Uncle Remus-like character, Brother Dickey, to tell the tall tale of having lit his pipe with a comet's tail:

FABLE AND PHILOSOPHY
HE LIT HIS PIPE WITH A COMET'S TAIL

"Dar's no comet in de big, round sky what kin put dis ol' worl' out o' business," said Brother Dickey, as he adjusted his big, brass-rimmed spectables, "an' you kin put dat in yo' pipe an' smoke it, wid my compliments.

"I'm older than dis ol' brick meetin' house, which wuz here long befo' de Civil war ever thought 'bout breaking' out, an' older than dat big oak tree out yonder, wid de long, gray ha'r hangin' from de top of its head, an' I has seen more comets than de arithmetic could count in ten nights, hand-runnin', an' ef I don't know what I'm talkin' 'bout no man in de worl' does. I was here when de stars fell an' folks thought de Day o' Jedgement had come, but dey didn't scare me!

"No, dey didn't. De only thing dat I was 'fraid of wuz dat dey'd burn up my watermelon patch, so I brought my ol' mule out de stable an' dat mule kicked 'em back ter whar dey come from as fast as dey could come in reach of his heels, an' when I sat out in de open, readin' a book by de light de fallin' stars gave, de folks dat saw me had confidence dat de worl' wuzn't comin' ter a end at dat pertickler time, an' some of 'em quit prayin' an' went a-swearin', an' yit dar wuz some dat knowed I wuz such a righteous man dey thought I'd decided ter go ter glory in a chariot of fire, an' axed me to take 'em along wid me, if dar wuz standin' room in de chariot, an' I didn't think they'd git burnt up on de way!

"Then when de biggest comet de worl' ever seen come along, wid a tail that stretched from one end o' de sky ter de other, till de night wuz as bright as day, dey thought that I wuz gwine to ride dat tail to glory, an' put in der applications ter let 'em sit behind me an' hold onto me till dey got dar! Why, I didn't do a thing but stand up befo' 'em an' light my ol' pipe wid de tip end of dat comet's tail—I sho' did!

"Dar's nutin' in de roun' worl' like being a righteous, self-collected man. But some of you didn't start soon enough on dat road, and I'm 'fraid yo' chance is slim."

Frank L. Stanton,
New York American,
May 22, 1910,

The May 18th issue of the comic magazine *Puck* featured this L. M. Glackens' portrayal of the comet as a Pullman porter. The inspiration for this piece undoubtedly lay in the fact the Chinese called comets "broom stars," the name derived from the shape of comets' tails. This was frequently mentioned in the popular press of the day. "Broom" must have suggested "porter" to Glackens, and the artist had a cartoon.

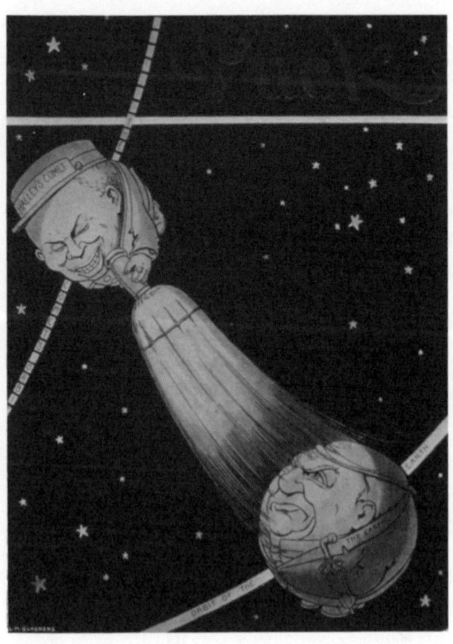

"Our Swede Servant" was a light-hearted column by William F. Kirk ("Little Bobbie's Pa") touching on the foibles of foreigners of the servant class. On May 17th Steena's neck was stiff after she looked for the comet from a rooftop with her fiance, a young policeman. The conversation between "our Swede servant" and the lady of the house contained interchanges like these:

"Ay ban op on roof last night looking for dis har Halley comet. Comets ban funny tengs. Ay keep looking op in air bout ten hours, it seem lak, and Ay never getting flash of comet at all."

"Do you know what a comet is, Steena?" I asked.

"Sure Ay know," said Steena. "A comet ban yust the same as a period, only it got a tail."

"You mean a comma is the same as a period, not a comet, Steena," said the wife.

"Ay mean yust vat Ay say," insisted Steena. "Ay mean a comet and Ay know a comet got a tail. My poleece faller tal me he see dis har comet right over Brooklyn last night, and he say end of vorld ban coming purty sune, so he asks me ef Ay skol marrying him yust before last payday he skol ever have.

"Sometimes Ay tenk it ban all a bunk," she continued. "It ban hard to tenk dis har beautiful earth ban going into a fire. Ay never used to believing in hal, and Ay hate to start now.

"Over in my church last night, before Ay go op on roof to see the comet, deacon faller ban making a prayer, and he say he know end of vorld ban coming sune. He say all dese sheep skol be parted from goats, and he say comet skol smashing earth von hal of a bump, only he ant saying it dat vay.

"Anyho, Ay got awful stiff neck."

The lady of the house plays the straight-lady to Steena's unintentional humor. This simple vignette must have tickled many a funny bone on the morning of May 17th.

Gentle, silly, bigoted, stereotypical, ethnic humor contributed its vision to an appreciation of Halley's comet's visit in 1910.

Sports Humor

As if to signify the kinship between humor and sports, every sports column that mentioned Halley's comet used it in a joke. William F. Kirk, who really hit comic pay-dirt in Halley's comet, wrote this little piece of sports humor.

LOOKS LIKE HALLEY'S
COMET IS TO BLAME

What makes St. Louis beat the Giants?
   The comet.
What makes the Pirates lose their science?
   The comet.
What makes our Matty get the gate?
Three Fingered Brown has lost five straight—
It isn't luck, it isn't Fate—
   The comet!

The baseball situation in the two big leagues at present lends color to the statement recently made by several eminent astronomers and scientists—the assertion that the proximity of Halley's comet is bound to have its effect on the denizens of this good old globe.

Some of the learned savants claim that the moment we get within hailing distance of the gaseous tail it will spell insanity and disaster to us all. They believe that it will be more or less a topsy turvy land; that wives will beat their husbands, instead of husbands beating their wives; that creditors will rush to hand their debtors more money; that Jeff will grow dark and Johnson white; that actors will buy all the drinks, and other strange phenomena.

We don't know that all this will come about, and, naturally, we hope that the presence of the strange visitor from realms afar will have no more effect on us than a May rain, but certain it is that in the baseball world everything is against the regular "dope" and has been for the last week or so. In the National League, for instance, humble St. Louis not only batted the peerless Matty out of the box in one awful inning and won again Saturday, but won both games in such decisive fashion that there was no room for a squabble. And didn't the domesticated Doves beat Chicago? And didn't Brooklyn trounce Pittsburgh? Granting that baseball is a charmingly uncertain game, the fact remains that something mysterious is coming off, and we might as well blame it on the comet. Kid Halley's protege can't come and go too swiftly to suit us.

This is stretching a thin tale to the point of disappearing.

The Jeff(ries) who will grow dark and the Johnson who will grow white appeared in a comic article promoting a *Denver News-Times* popularity contest! First prize in the contest was seats at the Jeffries-Johnson fight in July in San Francisco, accommodations and round-trip train tickets thrown in.

COMET'S LOST TAIL IS CAPTURED
BY NEWS—TIMES POPULARITY CONTEST

**Astronomers Are Again Peeved in Error, As
Instead of Becoming Less, It Is Growing
Longer and Bigger Each Day.**

The News-Times has found the lost tail of Halley's comet. Not only found it, but captured it and printed it in The News. That lost tail is right at the end of this article and since The News has discovered it, the tail is growing larger every day, instead of smaller as those bewildered astronomers would have you believe.

Having grown tired of sailing through the skies the tail dropped to earth. The News-Times captured it by the help of an agile reporter and is now running the tail every day. For the tail, gentle reader and readeress, is the growing candidate list that is run in connection with The News-Times popularity contest for those eight free trips to see the Jeffries-Johnson championship match.

If some people are too literary, why, the number of votes being garnered by the candidates in the News-Times popularity contest is growing faster than the comet's tail, anyway. Dating from yesterday there is one more month to go before the end, time for any ambitious person to win.

By May 21, over 970,000 votes had been cast for Denver citizens, F. W. Paget, Albany Hotel, having received 93,083 votes and for out-of-Denver folks 1,086,467, James K. Dye of LaJuanta having received 83,351 votes so far. But as the article said, there was time for any ambitious person to win.

The farthest-fetched case of humor was another fight reported in the *Ottawa Evening Journal,* May 19, 1910.

SPORTING
(Advance Wireless from Aurora Borealis.)

The set-to between Halley's Comet and the Earth was pulled off yesterday in the Ring of Saturn.

After a vigorous protest from Mars, Jupiter was appointed Referee.

The Man in the Moon occupied a stage box, and Venus appeared in a most recherche and bewitching apogee.

Owing to a wash-out on the Milky Way special trains from Uranus and Neptune were thirty years late, but the excursionists will probably be in time for the next mill.

The Comet shied his R. J. Devlin Company's Castor into the Orbit at 8.5.

The Comet led with his left for the Tropic of Cancer, but the Earth side stepped and the punch landed below the equator.

A foul was claimed but not allowed.

In the 2nd round the Earth jabbed Halley with the signs of the zodiac, and made him spit aerolites for several seconds.

First Cyanogen for the Earth.

A gang of roughs from Orion, who had been betting on the Comet, broke into the Ring, but the Referee got out some thunderbolts and quelled the riot in short order.

* * *

We regret to say that, owing to atmospheric disturbances of a serious nature, the dispatch was here interrupted, and nothing further, except a few disjointed phrases, such as "left hook to nucleus," "agony ended in 7th," etc., were received.

From this we judge, however, that the fight is over, and the vanquished Comet has retired—with his tail in the place it should be.

P.S.—Closed all day Friday.   The R. J. Devlin Co., Limited

And with the comet unconscious on the mat and the earth raising its arms in victory, it is time to leave the cartoonists and humorists of 1910, thanking them for the smiles and silliness they contributed on the occasion of Halley's visit.

18. "THE COMET PLAIN AND THE COMET FANCY": HALLEY'S COMET POST CARDS

The *American Stationer,* journal to the wholesale and retail stationery trade all over the United States, carried a regular column called "Post Card World." In the May 28, 1910 issue, in the midst of business news—"The plain sepia and sepia hand colored cards are new styles which E. C. Kropp & Company, Milwaukee, Wisconsin, are just commencing to make and in the estimation of the manufacturers are as good as can be imported from Germany," and an announcement of "…an elaborate line of souvenir postal cards…" issued by the Argentine Central Railroad Company—is this little fictional vignette:

Comet on Post Card

"Ho!" said a stout young man sitting in a cross seat on the elevated in New York.

"Hay!" said another young man, a friend who was sitting alongside. "What's biting you?"

"Nothing's biting me," said the stout young man, "but I've just been struck by an idea, and it makes me sick to think that I didn't think of it before. Here's this comet been careening around here for two or three weeks and no comet postal cards on the market. Think of the lost opportunities! Think of the cards we might have had with the comet just plain and the comet fancy and frisky, and think of the comet comic on postal cards that we might have had. Here's something that all creation knows about and is talking about and is interested in. Postal card people actually lie awake nights trying to think of new subjects for postal cards, things that will take, that people will buy; trying to invent things to feed this universal craze for postal cards, and here comes along right under our eyes the very biggest thing of all, and to the best of my knowledge and belief there hasn't been put out about it so much as a single card!"

Unbeknownst to the stout young man thinking of the lost opportunities, there *were* Halley's comet post cards produced in the United States and Europe, probably dozens. Post cards commemorating, spoofing, joking about, or announcing the comet were popular during the first six months of 1910.

By 1910 all major predictable natural phenomena from snowstorms to flowers to waterfalls were subjects of picture post cards. The simplest Halley's comet post card listed the best times to view the comet on the right-hand two-thirds of the card and included a drawing of the comet against a background of sparkling stars on the left third.

"It's only a paper moon,
Hanging over a cardboard sky,
But it wouldn't be make-believe
if you believed in me."

This song's sentiment and verse travel back through the history of photographic post cards to three earlier popular songs—"Moonbeams," (1906), "Shine on Harvest Moon," (1908) and "By the Light of the Silvery Moon," (1909)—for its beginnings and include a side trip to Halley's comet. The craze for foregrounds of all sorts of objects and quaint or odd backdrops lasted from about 1905-1920. Paper moons were a favorite, frequently selected and especially appreciated by those in love. For a few brief months there was an interest in comet backgrounds and even a foreground. Slightly embellished with a crudely painted Halley's comet and its name, this backdrop immortalized the subject's interest in the celestial event of the day.

A step beyond this innocent card was a hand-colored post card featuring a crude studio prop of the comet for the subject to straddle as this young man is doing. Vaguely suggestive props or pictures had a way of opening the channel for an equally suggestive note from the sender:

Dear Sam,

20 Sept. '10…I am still on the P.R.R. and still have the same old game at home you know what. I suppose you are keeping steady and preparing for the big day. I remain, Your old pal Jack Bill

It would appear that Sam was getting married. What "the same old game at home" was we can only speculate.

These cards exemplify the absurdly grotesque humor of post cards that profoundly inspired a later generation of artists. Surrealists Kurt Schwitters and Max Ernst freely acknowledged the contribution of this sort of post card art to their work.

The humor of these two post cards is wordless and simple.

Four simple cards wordlessly tell the story of the star-like comet's arrival, observed by their celestial friends, the sun and moon. Watched for with apprehension, the earth's worst fears are fulfilled when the star-like comet impales itself in the earth's face. But all turns out well in this series of cards. The figure of the comet as a tail behind a pointed star appeared frequently, often given the comet a dog's face.

As it turned out, the opportunity envisioned by the *American Stationer*'s stout young man—"Here comes something that all creation knows about and is talking about and is interested in"—was taken by some post card makers, and they, by gosh, profited from it.

19. HALLEY'S COMET AND THE AD-MEN

As could be expected, many advertisers hitched their wagons to the comet's tail in hopes of a lucrative ride. Shirt collars, eye glasses, soaps, perfumes, throat lozenges, printing services, champagne, bath powder, safe deposit vaults—what an unruly array of products gathered together beneath the comet's glow. There must have been great joy in the companies and advertising firms who concocted tie-ins with this attention-grabbing marvel of the skies. It is doubtful that any product got much more of the market than it could usually expect, but then that was not really the purpose. This sort of association would do any product good, and certainly no harm.

Quite a number of products for health, cleanliness, and beauty featured Halley's comet in ads appearing from April through June. "In the wake of MENNEN'S" babies bathe in comet dust, soft as MENNEN'S Borated Talcum. Pastilles Gerandel soothe the throat and cleanse the breath.

As a public service, Sapolio Soap included scientific information about Halley's comet under this ad:

Beneath a cartoon of a man reading from an eye chart for the oculist is this ad, tying Halley's comet in with eye glasses:

> You can see Halley's comet clearly with the naked eye if your eyesight is perfect. If, however, the comet is not clear and distinct, wear *Harris Glasses*. They enable you to see *just as nature intended,*

said M. H. Harris Oculists and Opticians of New York City. Since the comet was described everywhere as hazy to the sight and possessing a tail that resembled "gauze" or "hair," it is likely that the oculist led a few folks to believe that they needed glasses.

An English perfumer regularly advertised Shem-el-Nessim, "The Scent of Araby," associating the Mid-East and the heavens with his product. So it was just a simple shift of the imagination to substitute the comet for the moon and end up with the advertisement on the next page, right.

The advertiser's words almost mock themselves, since Halley's comet is with us again and Shem-el-Nessim is no more.

74

Wm. D. Perkins & Co., Bankers in Seattle, Washington advertised safe deposit vaults taking this line:

HALLEY'S COMET

Like other comets, according to astronomers, is a thing of mystery.

But there is no mystery about our

SAFE DEPOSIT VAULTS.

It is a cold hard fact that they are absolutely fire and burglar proof.

Place your valuable papers, jewels, and silverware in one of our boxes and cease worrying about a possible loss by fire or burglars.

Most of the companies and products have faded from the market place. Newmark's coffee, Thompson's furniture, Silver collars, Poinsetta soda—

Halley's comet has been burnt out by constant repetition …POINSETTA is just making its first appearance—new, spark-

are no more.

But surprisingly two products, Moët & Chandon champagne and Bird's Custard, are still with us to remind us of the comet's long life and of products which have been with us for several generations, beginning business in a vastly different world.

20. HALLEY'S COMET IN SONG AND MUSIC: OR, "I THINK THE THOUGHTS OF HALLEY, WERE ONLY A SWEET, SWEET DREAM"

In 1910 neither the phonograph nor the radio was yet in common use. They were brand-new ideas, each a fresh brain-child that was just catching on. Radio was just becoming able to transmit music and the human voice; up to that time it had only been used to transmit Morse code. So people heard and learned the songs and music of the day in vaudeville and dance and entertainment halls where they also bought sheet music to play at home. Local bands and orchestras were always on the lookout for the latest rage. It was almost inevitable that some composers would turn to Halley's comet for inspiration or seize the opportunity and stick the comet's name and image on their music as an attraction, a novelty, an advertising gimmick for musicians and audiences. The comet's name was an attention-getter itself, and that was what the composers and music publishers were seeking. In these rags, two-steps, waltzes, marches, schottisches, and polkas no sound particularly suggests a comet. Instead, the music is pure dance and "revue" numbers—songs and music—in the style popular in 1910. The comic material tickled a shallow funny bone, the gem of the Halley's comet songs appearing in Flo Ziegfeld, Jr.'s *Follies of 1910*. The highly popular Gus Edwards ("School Days," "By the Light of the Silvery Moon," "In My Merry Oldsmobile") composed "The Comet and the Earth" to words by Harry B. Smith ("Gypsy Love Song," "Shiek of Araby," "Yours Is My Heart Alone") for the most famous "girlie" show ever to grace the American stage, *Ziegfeld Follies.* All the well-known qualities of the Follies are here: the coy, shy, cute girl, the young man in love, and the kiss and sweet parting.

> *Follies of 1910,* which opened June 20, was probably the most extraordinary variety show seen up to that time in America. It began with a view of the Manhattan skyline and a replica of the New York Theater itself. A rehearsal was shown in progress ....Harry B. Smith made much playful fun of the techniques of presenting the *Follies.* In a later scene Anna was seen (an effect 25 years ahead of its time) in a film as a comet, with her smiling face emerging from a backdrop of stars. Suddenly she burst through the screen, mounted on a silver rocket, flew earthward, kissed Earth (Harry Edwards), and sailed back again.
>
> *Ziegfeld* by Charles Higham

The sentimental lyrics and verse of Cleo McCanse's "Halley's Long Looked for Comet" end on a chorus in which the singer first sees the comet and *then* wonders "how bright it will beam."

> **"Halley's Long Looked for Comet"**
>
> I think the thoughts of Halley, were only a sweet, sweet dream,
> And as he dreamed his mind was nearer to heavenly thoughts unseen.
> He awoke with great surprise, Thought himself nearer the skies—
> When mist rolled away, the sun filled his eyes, he declared the comet was no wonder.
>
> CHORUS
> The comet, the comet, the comet, I've seen,
> I wonder how bright it will beam,
> The comet, the comet, the comet, I've seen.

By yesterday's standards, the verses of the sentimental, inspirational songs stirred the emotions. There must have been many a home, even in Cleo McCanse's own Spring Hill, Kansas, where pianos played the very notes and someone sang these lyrics, and listeners and musicians alike were inspired by song.

Halley's comet music appeared on musical programs to the delight of dancers and audiences in entertainment halls throughout America.

Halley's Comet Sheet Music Covers: Images to Arrest the Eye and Stir the Imagination

Just as interesting as the music were the few extravagant Halley's comet sheet music covers produced to catch the customers' eye.

Bold print and a large comet with RAG in its tail make for a jazzy piece and complements the upbeat, bouncy cartoon guy and gal and bubbly heavens.

Similar elements appear in a staid and stately depiction of a Merlin-like astronomer studying the comet while standing on an earth dramatically reduced in size.

This fanciful sheet music cover for "The Comet March and Two-Step" pictures the comet faintly visible behind an Arab astride a camel. The cover seems to be saying, "An oasis, a camel, Halley's comet and you." But why? Lost to memory is the reason for the publisher's choice of image to represent William Leander Sheetz's lively "Comet March and Two-Step."

21. A COMET SERMON FOR YOUNG FOLK

A sermon for young people appeared in *The Christian Work and Evangelist*. Taking as his text "Wandering Stars" author Rev. G. B. F. Hallock, D.D., presents many of the widely known facts about Halley's comet, then launches a wonderful series of similitudes for children.

The Bible takes up this thought about comets, or "wandering stars," and applies it to certain kinds of people. Let us trace some of the features of similarity.

I. In the first place, some folks are very much like comets, in that there is not much substance to them. Like with comets so with some people, the biggest part of them is gas. Some are big enough, and showy and pretentious, but there is not much to them.

1. Now, the first thing wanted in young people is a sound and solid body, as much as possible unlike the thin, vapory substance of a comet. Physical strength is one of God's good and perfect gifts, and is an adornment and glory for anyone to possess. Temperance and exercise are the best doctors. Never encourage that very common and mistaken idea that good people must be sickly and delicate and look heavenly, that religion is intended more especially for the weak. Instead of that, religion makes a special claim upon the strong and vigorous because they are strong. Young people, care for your bodies, cultivate them, keep them pure and clean, as far as possible, make them strong, and then consecrate them and all their powers to Christ.

2. The second thing wanted in young people is a good and solid head. Some comets are peculiar in that they have no head. But some people are also peculiar in that respect, too. It is good to have a sound mind in a sound body. It is a good thing to have a head. Have a mind and use it. It has been well said that there are two reasons why some people don't mind their own business: one is that they haven't any business, and the other is that they haven't any mind. It is a good thing to have a mind and to use it. A business man was asked the other day how he spent his time. He replied, "In the daytime I mind my store and in the evening I store my mind." Use, cultivate, improve all your mental powers. Bring them to bear upon the great subject of religion. Work out the problem, that great Bible problem, "What shall it profit a man if he shall gain the whole world and lose his own soul?" If you have worked out that life insurance is profitable, see also what you must think of soul insurance. Is godliness profitable, will it pay?

II. Notice, in the second place, that some people are like comets in that they are easily swayed out of their orbits. The metaphor applies to unstable men, driven hither and thither by temptations, whose life presents the strongest kind of contrast to the safe, well ordered life of Christians, more fixed, like the orbit of a planet. Comets are often swayed out of their orbits, planets never. In 1776 Lexell's comet was seen and then lost. The whole world wondered for a time, but now all is plain. It had come so near to the planet Jupiter that the attraction drew it out of its orbit.

1. Young friends, keep to your orbit of purpose. Have an aim and stick to it. There is power in a purpose, a presiding purpose. That life that is not ruled by it is only existence. It lacks the elements of life, aim and action: it lacks the traits of life, earnestness and energy; it will lack the end of life, a sure success. Many a life goes to waste and ruin simply because, like an abandoned and drifting vessel, or a wandering star, no guiding purpose directs its course.

2. More important, young friends, keep to your orbit of right. Let no Jupiter attraction sway you out of it. "Dare to do right, dare to be true." Decision of character is certainly a noble quality. It is a glory to any man to be able to say, "No," and mean "No," or to say, "I will," and, like Daniel, to mean, "I will," in spite of the lions. Remember that if you lack moral backbone you might just as well be a jellyfish, so far as accomplishing any good in the world is concerned. True manhood has a far nobler ambition than that. "Trust in God and do the right."

III. Notice, again, that comets grow brighter as they get near the sun and darker as they go away from it. So do we all grow more bright and beautiful as we get near to Christ, the Christian's Sun, and darker as we go from him. The brightest and best and happiest men and women you know are men and women who live near to Christ. The most unlovely and repulsive are those who live far from him. Like comets we get all our light from our sun, we shine only with reflected light; let us keep near him, that we may reflect his life and character.

IV. Notice, lastly, that some comets are truly "wandering stars." As unstable, disrupted ruins, they are hastening forward to a final darkness among the slags of the last process of reconstruction. Surely this is very suggestive of the sad ending of sin. "The wages of sin is death." Sinners unrepentant are "wandering stars, to whom is reserved the blackness of darkness forever." To die in one's sins is the darkest of deaths. It is to make a failure and worse than a failure of life. But there is no need to make a failure, for "God so loved the world that he gave his only begotten Son, that whosoever believeth on him may not perish, but have everlasting life." The thought in your mind is to make sure by becoming a Christian now. Do it. It is the prompting of all that is highest and best in your nature and is the only real wisdom.

Rev. G. B. F. Hallock, D.D.

V
The Crack of Doom

22. PORTENTS, PROPHESY, AND EDWIN EMERSON'S BOOK OF TERROR

> *Those rare beings who have lying latent within them the gift of Second Sight or divination.... upon the near approach of Comets find themselves stirred to prophesy. Hence, so many marvelous prophesies inspired by Comets since the ancient days of Merlin, the seer.*
>
> Edwin Emerson, *Comet Lore,* 1910

Because they *seem* to come from nowhere and obey no apparent regular laws of the heavenly bodies, comets—and above all Halley's—were regarded as portents of coming disaster or warnings in times past. As Edwin Emerson wrote, "In the train of Comets, it has ever been held, come wars, bloodshed, fire, floods, plagues, famine and the fall of mighty rulers." He reminds us:

> Our Holy Bible confirms this time-honoured belief. The Savior Himself said, according to the Gospel of St. Luke, Chapter XXI, Verse 10-11:
>
> Nation shall rise against nation, and kingdom against kingdom; and earthquakes shall be in diverse places, and famines and pestilences; and fearful sights and great signs shall there be from Heaven.

In the Revelation of St. John the Divine (Chapter XII, Verse 14) we read:

> There was seen another sign in Heaven, and behold a great red dragon...and his tail draweth a third part of the stars in Heaven. And behold the third woe cometh quickly.

In the tradition of the Biblical prophets and God Himself, even in 1910, a truly enlightened age regarding knowledge of the true nature of comets and other heavenly bodies, "prophets" spread their dire predictions and "yellow" journalists and pseudo-scientists incited fear.

The comet was simply an object taking part in an ordinary sequence of astronomical events. Halley's comet was nonetheless seen by the prophets as a sign, in human and divine history. Dire predictions appeared in leaflets and pamphlets. Basing their prophesies on Biblical injunctions, numerology, astrology, and other forms of divination, the soothsayers foretold the apocalypse, the end of time, the doom and doomsday of the world. The worst (or best) of these fear-inspiring pamphlets was Edwin Emerson's *Comet Lore.* Such chapter titles as "The Terror of the Comet," "Great Events and Disasters Linked with Comets," "Halley's Comet the Bloodiest of All," "Halley's Baleful Comet," "This Year's Prophesies," "Our Peril from Collision with the Comet," and "The End of the World" indicate Emerson's tone. As a rhetorician Emerson is a master of "sleight of thought." He is slippery as an eel, and we will watch him wriggle through one exercise in persuasion.

Emerson's unique contribution to the lore and literature of Halley's comet's 1910 apparition was his discussion of prophesies regarding the comet's impending influence on human affairs. In the chapter "This Year's Prophesies," Emerson reported that in December 1909, the French astrologer and prophetess, Madame de Thebes, predicted the coming of a comet early in 1910 to be followed shortly thereafter by disasterous floods in Europe. In January, Inness' Comet or "1910, A," a comet never before seen and of short duration, appeared in the heavens. And as Madame de Thebes predicted, "The disappearance of the comet in France was followed by unprecedented rains and floods which covered one-fourth of France...inundated Paris, completely submerging all the bridges over the Seine." There were also floods in Germany and Italy. Then, on January 20, 1910, Madame de Thebes uttered prophesies specifically about Halley's comet's effects upon the world and human affairs.

> This year, 1910, will be one to look back to with trembling.
> The earth is under a terrific strain from Comets and planetary revolutions. Human destiny is red. That means blood. Political events are black. Terrible changes are imminent....
> The coming of a Comet will affect us for the worse.
> The strain of the stars will be most severely felt in America. The people of America will have to pay dearly for all their riches and sudden prosperity. With the coming of another Comet disaster will descend upon America.
> A financial crash is impending.

Madame de Thebes also wrote:

> There will be a long string of suicides. Black ruling us, men will commit all manner of crimes and knaveries for money. The times are swinging toward degeneration. We are swinging within the evil influence of a strange orbit. Our souls are jarred from their proper bearings. I dare not say all that is revealed to me. It would be too terrible.

Following this report of Madame de Thebes' utterances, Emerson cites two other comments by highly placed persons, *implying* that they also see the comet as the cause of impending disaster.

> Before Madame de Thebes' ominous prophesy concerning Halley's Comet and its effects upon America were cabled over to this country, another, no less dire prediction of financial disaster in the United States, coincident with the appearance of Halley's Comet, was made by W. E. Corey, the President of the American Steel Trust. Mr. Corey warned his friends to "call in their money and get from under" because a calamitous financial crash and general business ruin would surely come during the Spring of 1910.
> The most ominous of all prophesies connected with the coming of Halley's Comet this year was made by the venerable General William Booth, the head of the Salvation Army. Speaking in London, immediately after Halley's Comet had been located early this year, General Booth said:

> "We are, this year, rapidly approaching the end of all
> things, with similar results, but far surpassing in horrors any
> disaster that has gone before.
> "All things will be wound up. Besides a deluge of water
> sweeping part of the world and its inhabitants there will be fierce
> destruction by fire."

The fact is, neither Mr. Corey nor General Booth *mentioned* Halley's comet: Emerson creates the *impression* that these gentlemen claimed that the comet was the cause of the disasters they predicted.

Emerson's intention was not to allay people's fears—as other writers were striving to do—but to incite them. While Madame de Thebes' predictions are presented in a sraightforward way with allowable vagueness proper to the art of prognostication, Emerson's own words, on the other hand, are cleverly deceptive. Innuendo, implication, hyperbole, or overstatement and a wide range of rhetorical tricks create a picture of the outcome of Halley's comet's apparition far different from that which legitimate science and journalism presented.

23. HALLEY'S COMET IN LURID ILLUSTRATIONS AND DRAMATIZATIONS: "A FASCINATING EXERCISE FOR THE IMAGINATION"

With portrayals of collision between the comet and the earth, the earth reduced to rubble and ashes, massive tidal waves, lethal poisoning of the atmosphere, the "yellow" journalists and illustrators incited the fears of a gullible and ignorant public.

Halley's comet attracted a species of writing characterized by a tone of "science reporting" which masked the intention of scaring the living daylights out of the reader. Qualifying this sort of writing as "exercises of the imagination," as one writer called his essay, and the working out of hypothetical possibilities (hypotheses rejected by professional astronomers) for Halley's comet's 1910 apparition, are lost to the unsuspecting reader who, given the anatomy of the mind, dwells in fright on vivid depictions of a horrible end to life. "The Danger of the Comet" by E. C. Andrews, lavishly illustrated by "Keok-Keok," appeared in the December 1909 issue of *Pearson's Magazine* (London, England) and was censured by the scientific community. Andrews mixed accurate scientific and historical data with speculations about what would happen "if."

> Is it conceivable that a comet will one day collide with the
> earth? Collisions have taken place between comets and stars,
> and the alarming possibility of such a collision affords a fascinat-
> ing exercise for the imagination. Halley's comet, which appears
> at roughly 75-year intervals, is now almost due, and has actually
> been sighted by astronomers in the infinite distance of space.
> The results upon earth of a collision with such a body, travelling
> at the rate of some million miles an hour, are here vividly de-
> picted by word and picture.

"The Danger of the Comet" begins with an overblown description:

> Hundreds of thousands of miles of white-hot matter, foam-
> ing through space like a vast cataract, and travelling towards
> our world at a million miles an hour—such is Halley's comet, the
> coming comet of 1910.

Following these frightening exaggerations, Andrews recites the facts of this and earlier apparitions of Halley's comet using a tone of historical objectivity. This tone or pose tends to make the wildly speculative observations of the article appear to be factual and objective, too, thus feeding the dread readers must have felt. The facts make the fraudulent and excessive passages believable. The movement back and forth between factual reporting and fear-mongering characterizes the type of article Andrews' piece exemplifies. There were a dozen or more similar treatments in as many popular magazines and newspapers. The writers wrote as if they were unaware of the fear their articles might engender.

The effects of "The Danger of the Comet" are heightened by vivid depictions of the worst scenarios Andrews concocted.

The first illustration shows an immense comet about to strike a city already showing signs of devastation. Fires burn, people are dead or writhing or trying to escape the imminent catastrophe while buildings are collapsing. Ghastly shades of purple and orange enhance the effect of this terrifying image. The caption briefly recites the ultimate catastrophe which awaits mankind:

If a large comet approached within measurable distance
of the earth, the doom of our world would be sealed. Such
tremendous heat would be engendered that everything would
spring into spontaneous combustion. The hardest rocks would
become molten and no living thing would remain upon the
earth's surface. Buildings and human beings would be
scorched to cinders in a second.

Later in the article the effects of a less horrible event of "…the comet sweep[ing] the outer fringes of the earth's atmosphere…" are dramatized in a frightening seascape. Ships are utterly destroyed and buildings wrecked by "…the vast tornado that the passage of the comet would cause. It would…probably devastate one complete hemisphere."

The scientific community was outraged and appalled by such outright fear-mongering. Addressing a meeting of the British Astronomical Association on December 29, 1909, Dr. Crommelin, the astronomer who calculated the date upon which the comet would reach perihelion, said that this "…sensational article…was a revival of the worst type of comet panic." The world had seen others in relatively recent times, all the result of this sort of writing.

Such writing also tended to discredit legitimate science, since the articles often quoted "professors" and used "scientific" language and "reasoning" while distorting the truth about Halley's comet.

A drawing that depicted the comet about to strike the earth and the tidal wave its approach produced about to break upon a devastated town appeared in the *Sketch* for April 27, 1910 under the headline: IF YOU FEAR THE COMET, LIVE ON A MOUNTAIN TOP. The advice was both too late and unnecessary, but appearing just a few weeks before the earth and the comet reached their closest proximity, it is easy to imagine the horror such illustrations must have inspired. While the caption and illustration proposed to allay the fear of the readership, the overwhelming force of both was to frighten the less informed and more gullible of the readers. The subjunctive, "Should the body of Halley's Comet strike the earth…—which we may assure our readers is not in the least likely…" assures the reader that the event will not take place. At the same time the caption vividly and convincingly portrays the earth's ruin:

IF THE HEAD OF THE COMET BUTTED MOTHER
EARTH: THE SEA RISING IN BOILING WAVES
AND OVERWHELMING THE WORLD.

Should the body of Halley's comet strike the earth—which
we may assure our readers is not in the least likely—there is no
doubt that an immense catastrophe would follow. Some have
argued, even, that the globe would be reduced to its original
chaos. Indeed, several perturbing predictions have been made.
Laplace claimed, for instance, that on the approach of the comet
tidal waves of boiling water, thirteen or fourteen thousand feet
high, would rise, leaving only the higher Himalayas and the Alps
above water. If you believe Laplace, therefore, make your habita-
tion on a mountain-top.

IF YOU FEAR THE COMET, LIVE ON A MOUNTAIN-TOP.

    With this dramatic picture we leave this "fascinating exercise for the imagi-
nation." Collisions between Halley's comet and the earth, the earth reduced to
ash or "original chaos," massive tidal waves, lethal poisoning of the atmosphere
—the yellow illustrators and journalists must have enjoyed scaring the public,
much the way a harmless prankster enjoys his tricks.

VI
Events

24. "NOW I LAY ME DOWN TO SLEEP:" FEAR, PRAYER AND SUPPLICATION

Fear that Halley's comet would cause harm or disaster moved many all over the world to pray, begging their God to save them. Even as early as April 10, fully a month before the earth's predicted passage through the comet's tail, the *London Observer* reported events in southern Russia under the headline:

TERROR OF THE COMET
Prayer and Fraud in South Russia

the correspondent reported that:

> …prayers were being offered in country churches and monasteries for the salvation of Russia from threatened cataclysm. During the last week a paragraph vaguely purporting to represent the views of certain distinguished Russian astronomers has been going the round of the provincial journals to the effect that the visitation of the comet will have the most disastrous consequence.
>
> The superstitious fears of the peasantry are being exploited by persons who are collecting money, falsely representing that it is for the solemnization of masses and special services during the dreaded month of May.

As the date of the earth's passage through the coment's tail drew near, incidents of terror and the religious fervor accompanying it spread. For the ten days prior to May 18th, superstitious Mexicans sought to turn aside the impending disaster with music, chants, and ceremonies. Throughout the border area between the United States and Mexico they had erected crucifixes on the hills, around which they gathered each night and watched for the appearance of the fiery star. In pueblos and ciudads many gathered in churches and spent day and night in prayer. Hundreds hid in caves and canyons in the mountains until the comet passed. As the night of May 18th passed without catastrophe, the religious ceremonies turned into thankful dancing and feasting.

All over the globe people were terrified on May 18th, turning to prayer and supplication for salvation. A missionary by the name of Will Pattillo reported vivid and marvelous events in his neighborhood in Japan for the *Japan Weekly Mail*.

> …A great many people in my neighborhood fully expected that at noon on May 19th they were going to be crushed into atoms by a collision with the comet or be suffocated by its gases. On that day thousands of people quit work and spent the day calling upon their gods for protection, while thousands of others went on with their work in fear and trembling. One man committed suicide two days before, leaving word that he did so because if he waited till the catastrophe came everybody would be dead and he would have nobody to bury him. His selfishness was even greater than his foolhardiness. Hundreds of factory girls quit work and returned to their homes, to die with their parents as they said, a touching but pitiful tribute to filial love. Along a country road two men were seen carrying on their

shoulders a pole to which a box was attached containing Budd-
hist prayerbooks. Occasionally they stopped, and the farmers
came out of their fields and crawled on their knees beneath the
box. This was a talisman to ward off any harm from the comet.
Others carried rice dumplings to the temple and after offering
them up to the god, a portion was eaten as an antidote, so to
speak.

In Manila, the Phillipine natives' fear of the comet reached a climax on May
18th. Immediately after a large bolt of lightning and a loud clap of thunder struck,
the electric trolley cars stopped suddenly. Many of the occupants of the cars,
recalling the presence of the comet, knelt in the cars and prayed for safety.

Peasants in a small Hungarian village believed that the world would collide
with Halley's comet and be destroyed. On the night of May 18th, seeing a blaze in
a nearby village, the village watchman went through the streets blowing a horn to
wake everyone. He told those who hurried out of doors that the world's end was
at hand. The women and children screamed with fright and the men, overcome
with terror, hurried into the houses and took out all the food they could find. They
carried the food to the square in front of the church where a great fire was set
and everyone feasted, gorging themselves on all the edibles they could find.
Between mouthfuls the villagers prayed.

Hundreds of Puerto Ricans paraded the streets of San Juan and other
towns in Puerto Rico on the 18th, carrying candles and chanting prayers. Many
of them also spent much time in confessionals. A large number of workmen
failed to show up at the tobacco factories and plantations, and the pineapple
shipments were curtailed because the laborers refused to work.

Fearing that the comet's tail was about to destroy the world, John Chiquala,
a Lummi Indian well-known in the city of Bellingham, Washington, went about the
streets with an open Bible, imploring people to be saved from destruction. Earlier
Chiquala had told members of the Lummi tribe that there was no use cultivating
the fields in the spring since the comet would destroy the earth.

In Seattle the sighting of the comet caused widespread fear among Orien-
tals and natives of the islands, and many were thrown into a panic when it
became known that the earth was to pass through the tail. Japanese laborers on
two large plantations refused to begin work in the morning and spent the day in
hiding; overseers all over the islands had great difficulty in keeping the men at
work.

Hundreds of Russians and Mexicans working in the fields of Fort Collins,
Colorado were greatly excited the night the comet was closest to the earth. They
believed that catastrophe was approaching. Children were kept home from
school during the day, and most of the time was devoted to prayer.

The *American Review of Reviews* reported, "Many Negroes of our Southern States, particularly, are reported to have been in a state of mind bordering on
frenzy or fear." The Negroes of Ashville, North Carolina were in a state of frenzy
believing that the end of the world was at hand and that when the earth passed
through the tail of Halley's comet time and space would be no more. Blacks held
religious services all over the city. Their prayers and pleadings could be heard for
several blocks. Standing room only was to be had at their churches. Crimes among
the Black population, according to police court records, had slumped remarkably

in the two days before the 18th of May. And it was said that Negroes couldn't be hired "for love or money," for they said there would be no more pay days. At Stanford, Kentucky scores of Negroes professed salvation at all night church services in order to prepare themselves for whatever might happen the next day.

Miners held conflicting views, some believing it safest to remain underground, while others wanted to perish on the earth's surface with their families. Several thousand miners in the anthracite coal region of Wilkes-Barre, Pennsylvania, chiefly foreigners, refused to enter the mines on May 18th. Efforts by the English-speaking miners to get them to go to work were futile, and the foreigners said that if the world came to an end they wanted to be on the surface where they could see, instead of in the depths of the mines. Many spent the day in prayer, greatly fearful and nervous. Some collieries were so short-handed that they had to shut down for the day. By a different line of reasoning, many foreign miners employed on the day shifts of the Leadville and Cripple Creek mines in Colorado stayed underground at night to avoid contact with the comet's tail. In Brazil, Indiana, many foreign coal miners drew all the money that was coming to them, lavishly spending their money that night on things that could be quickly enjoyed.

Johannesburg, South Africa reported that despite reassuring statements from the local observatory there was an extraordinary amount of nervousness on the Rand regarding the passage of the earth through the tail of Halley's comet. The wife of a prominent mine manager lived in a special chamber which she insisted her husband have built for her at the bottom of a mine until the passage through the comet's tail was over.

To many of the superstitious foreigners in America, May 18th was a trying day. In spite of reassurance from astronomers all over the country, published in the American and foreign language newspapers, hundreds of the foreign-born felt that something dreadful was going to happen. In several instances their pent-up anxiety burst forth. In the lower part of New York City where foreigners lived there were some scares. At Mulberry and Broome Streets twenty or so Italians were looking at the sky when they saw what appeared to be a ball of fire shoot out from behind some fleecy clouds. The on-lookers yelled with fright, several falling to their knees. Their shouts attracted the neighborhood and in a few minutes several hundred people were at the corner watching the light. Suddenly the ball burst and smaller balls of fire fell rapidly toward the watchers. Instantly many of them began praying for deliverance, some calling aloud to their patron saints. A few in the crowd, however, kept their nerve. They found that the little balls of fire, which their compatriots mistook for pieces of the comet, were really pieces of a burning toy balloon. They shouted to their companions, explaining what the fiery specks really were, but while they were still yelling the police arrived and dispersed the crowd.

Many foreigners living in Constable Hook, Bergen Hook, and Bergen Point were badly scared about the comet. Most of them refused to work on May 18th, and many spent the day offering prayers in church for deliverance from disaster. Every Italian employed by the Ten Broock, Morse and Eddy Companies of contractors laid down their tools in the morning, believing that the end of the world was about to arrive, and spent the day in church.

That same night from the four great bridges across the East River of New

York City, from the decks of ferry boats, from pier ends and wharves, from the islands in the harbor, from rooftops, and for the entire length of Riverside Drive, the people of Manhattan clustered to look for the incandescent tail of the comet and for the comet itself. Some prayed in fear. Word had been received late that day about violent disturbances observed on the sun. The news added greatly to the fear of those in the crowd who reasoned that if the sun were susceptible to the comet's influence, then the earth must be also.

So it was all over the world on the night of May 18th, a world in 1910 largely populated by an illiterate peasantry lacking any means to hear what science had to say.

An editorial appearing on May 19th in the *Seattle Post-Intelligencer* spoke about the effect the comet's passing had.

> The comet came, the comet went and this old earth is no worse and no better and thus far no wiser. There was no collision as the superstitious and the ignorant feared, and now that the comet is headed away from us, there will be no recurrence of the manifestations of terror that were recorded from all parts of the world.

Of course for most people the comet was simply a curiosity, a wonder of nature, something to see and enjoy and marvel at. And so the world lapsed into quiet again for the moment, people turning their fears to other terrors, real terrors —wars, civil disorders, human disasters of every sort, the stuff of every day's news. Halley's comet was a blessed relief, even to have fears turn out unfounded, as they so often do not.

25. HALLEY'S COMET AND
THE DOGS OF CONSTANTINOPLE

The events of May 18th in Constantinople first were told in the *London Times* and were reprinted in *The New York Times*. The *Times* headline put the story this way:

FEARED WORLD'S END IN
CONSTANTINOPLE

Many Inhabitants Expected Death When
the Comet Approached.

But Only Dogs Suffer

Authorities Chose that Night to Gather Them in, and Their
.   Fate is a Mystery—Made into Gloves, Perhaps.

The editors must have enjoyed it as a story to chuckle over, jovial news-room talk: "Did you hear the story about what happened to the dogs in Constantinople last night?"

In Constantinople, a city where every beast of burden and most children wore charms to guard against the evil eye, where astrologers told the future, and where eclipses were watched with terror, Halley's comet was watched by the horror-stricken. At twilight on May 18th, 100,000 frightened people came out on the roofs in night clothes and passed the night there; some to watch the comet, others to comfort the fearful, others to meet the terrors of the last day with friends and neighbors. Some prayed, others sang to music. When dawn came the natural order of the earth and the heavens was seen to be as it had always been.

The same night the municipal authorities, unaffected by celestial signs and portents, marshalled all their forces against the famous street dogs of Constantinople. All night, in the quarters of Pera and Stamboul, policemen and sweepers, armed with lassoes and huge wooden tongs, hoisted and tossed the squealing and yelping dogs into dust carts. What a strange noise it added to the night's terror. A few dogs were spared by owners who bailed them out. The owners had to swear to get collars with name tags for the dogs and to pay the municipal dog tax. The captured dogs were driven away, their destination uncertain. Some said they shared the fate of "reactionaries" and were marooned on a desolate island, and left to die. Others said their skins were made into soft gloves.

So passed the night of May 18, 1910 in Constantinople, harmless to mankind, doom to dogs.

26. HYSTERIA, MADNESS AND SUICIDE

Fear of annihilation
Leads Weak-Minded
To Suicide and Crime

(Seattle Post-Intelligencer)

The comet's approach drove some people harmlessly insane, and others developed suicidal mania.

Madness

Miss Kate VanNess of Carlton Hill, New Jersey was taken to the Morris Plains Insane Asylum on May 19th. Her mind became unhinged following a discussion of the comet's appearance. On the way to Morris Plains she said she would follow the comet no matter where it went.

That same day in New York City, Samuel Popowski declared that the tail of the comet was striking him all the time, indeed, it was beating him into shreds. City Physician Baldwin pronounced him insane.

The comet's tail also chased James Kline on May 22nd. Formerly a Pullman car porter, Kline was put in the Somerset county jail in Somerville, N.J.—a raving maniac as the result of waiting in terror for five days for the destruction of the world by Halley's comet. A policeman was standing on Main Street early in the morning, when Kline ran past in scant attire shouting that he was being pursued by his mother-in-law and the tail of Halley's comet. Religion played a part in many of the delusions people suffered. When the officer shouted, the fleeing Kline stopped suddenly and began to pray. Kline, sober and industrious, had only a week earlier gone through the Negro colony warning his brethren to prepare for the end of the world.

94

Brooding over the comet brought on an attack of religious frenzy in a woman living on the northwest side of New York City. She was on her way downtown in a street car, when she suddenly rose to her feet, shouting and gesticulating wildly. "Glory, glory, glory!" she cried. "Get down on your knees, you sinners, and pray the Lord to forgive you, for this is the end of the world." Then she sought to force men and women in the car to get down on the floor and pray. A policeman was called, and the woman was taken to the station, where she was unable to give her name.

By far the most bizarre manifestation of insanity combined with religious frenzy struck Paul Hammerton. The visit of comet 1910, A alarmed Hammerton, and when he learned that the earth was to pass through the tail of Halley's comet his mind gave way. He believed that the end of the world was at hand. While brooding over the comet's visit, the sheepman and prospector became insane and crucified himself, according to the mining men who brought him to San Bernadino, California on May 9th. Hammerton had nailed his feet and one hand to a rude cross which he had built. Although he was suffering intense agony, Hammerton pleaded with his rescuers to let him remain on his cross.

Suicides, Attempted and Successful

Mrs. Viola Gastenum of Anaheim, California tried to murder her two children and kill herself afterward. She gave her two children doses of concentrated lye and drank some herself. All three were near death when discovered. Mrs. Gastenum said the comet was sure to destroy the earth and everything with it. She wanted herself and her children to escape a fiery death.

Blanche Covington made up her mind that there was no escape from the comet and that it would kill everybody in Chicago. She had difficulty in convincing her friends, but this did not change her opinion. Dreading the suffering that she might have to undergo, she locked herself in a room and turned on the gas. Mrs. Marie Welch called a policeman and, with his aid, rescued Miss Covington.

Under the *Rocky Mountain News* headline,

FEAR OF COMET DRIVES
WOMAN TO SEEK DEATH,

appeared the story of Mrs. Jeanette Niebert's suicide. Helpless after weeks of illness and fearing that Halley's comet was going to annihilate the earth, Mrs. Niebert swallowed morphine on May 18th in Denver. That morning Mrs. Niebert told her husband Charles that she sensed impending danger, a feeling she could not free herself from. She said something would happen to her before the day ended. At noon she told friends that she wanted to die. She could not stand her helplessness. When her husband returned that afternoon she told him that the presence of some intangible danger made her helpless. When Charles arrived home at 6 o'clock Mrs. Niebert was sinking into unconsciousness. He immediately called their doctor who in turn notified the police. Mrs. Niebert did not recover. Her last words as she closed her eyes were: "I think—the comet!"

Suicides were committed in fear of Halley's comet, but Bessie Bradley, twenty-five years old, committed suicide on May 19th at Hastings-on-Hudson, where she was employed as a maid, because the comet *failed* to appear! The young woman became greatly worried over the contradictory reports of what the comet would do and what it would not do. When no sign of it showed on the night of May 19th she grew so nervous that she could not sleep, and in the morning she went to her room to rest. Other servants who went to wake her for lunch found her dead on her bed, the gas turned on.

Determination did not result in success for one man. On May 21st, W. J. Lord of Cottonwood, Alabama was in a precarious condition as the result of four attempts to commit suicide. With his mind wrought up over Halley's comet, and believing that he had sinned against the Holy Ghost, Lord had attempted to shoot himself. Unsuccessful in this, he jumped off a roof and fell on his head, knocking out his teeth and sustaining other injuries. He then cut his throat and jumped into a well.

There were probably dozens of suicides caused by fear of the comet's approach. The deaths were pointless as it turned out and reveal the depths of fear that some experienced. Perhaps those who failed to bring about their own deaths can be counted the most successful.

27. THE STOPPED HEART: HALLEY'S COMET AS A CAUSE OF DEATH

There were only six recorded instances of fatalities directly related to people looking for or seeing Halley's comet. Three children fell through skylights, and one was accidently shot while searching the sky for the comet on the night of May 18th. Three adults had fatal heart attacks when they saw the comet, one on the 18th and two on May 22nd.

Amy Hopkins, sixteen, attended a "comet party" with a dozen friends on the roof of a four-story apartment house in Brooklyn. Miss Hopkins was sitting on the skylight when the framework broke and she fell. Her skull was fractured and her neck was broken. Mrs. J. B. Hopkins, her mother, was on the roof of a house across the street. Hearing the screams of the young people, she hurried to the other building, and reached it just as her daughter's body was being taken from the bottom of the shaft. Although several of the young people from the party tried to keep her from entering, she pushed her way past them to see the lifeless body of her daughter. The ambulance surgeon, Dr. Griffin, of St. John's Hospital, said the girl had died instantly. Mrs. Hopkins fainted, and was in serious condition and put under a physician's care.

On East 110th Street, eight-year-old Louis Happer fell through the skylight of a five-story building while looking for the comet. He died of a fractured skull.

And there was at least one "close call." Seventeen-year-old Miss Anna Esmess of the Williamsburg section of New York City decided to join a "comet party" with her friends on the roof of an adjoining house. She was on the roof of her own apartment building at the time, and rather than go all the way downstairs, next door and up again, she tried to jump across the five-foot gap between the buildings. Missing, she fell three stories, bouncing from side to side in the airshaft, thus slowing the speed of her drop. Miss Esmess was knocked unconscious, but by the time a surgeon arrived from the Eastern District Hospital she was able to walk home.

A child of eight looking for the comet through a broken bottle—the best telescope she had—was shot during a street fight in Little Italy. Her spine was shattered and she died.

Three adults suffered fatal heart attacks, apparently as a result of the fright caused by looking at the comet. On the night of May 18th, Mrs. Willie Morris, a black woman from Fredericksburg, Virginia, was one of a crowd gathered on the bridge over the Rappahannock to watch for the comet. She expressed great alarm and then fell dead.

On the evening of May 22nd, the comet caused intense excitement in Taladega, Alabama. The congregations of several churches left their pews and hundreds of excited people stood in the town square gazing at the celestial visitor. Miss Ruth Jordan, daughter of a farmer living two miles outside of town, was called to the door of the family home to see the comet and immediately fell dead. Physicians claimed that heart failure was the cause.

The comet was pointed out to an unidentified black man on the Taladega depot platform, and he instantly dropped dead.

28. COMET PARTIES

Comet party on the roof of the Waldorf-Astoria.

New York City Hotel Parties

Four days before the tail of Halley's comet engulfed the earth, interest reached the guests of New York hotels. The roofs of all the big hotels were crowded nightly with comet parties. Reservations were being made at many of the hotels to watch the comet for each of the four nights up until May 18th. Many hotel guests selected their rooms so they could view the comet from the windows of their rooms.

Some of the hotels printed a time sheet showing the comet's expected appearance in the early morning hours. Not in the history of the hotel management had there been such a demand for wake-up calls at 3 o'clock in the morning.

Dr. Ralph M. Grace, a physician and amateur astronomer who was staying at the Hotel Gotham, invited a dozen of his friends to watch for the comet and then entertained them with music and breakfast. Part of the hotel's roof was arranged as a small garden, decorated with palms and Japanese lanterns suspended so that their colored lights would not interfere with the view of the comet. Side awnings were hung to keep out the cold. A round table decorated with flowers was spread for a 2:30 a.m. breakfast. The menu was planned so that all of the dishes reflected interest in the comet. The guests sat around the tables in their motor car toggery of fur coats and caps. A Hungarian orchestra played behind a screen of ferns and palms, one of their selections being "A Trip to Mars." The comet was sighted by the outlook just after the first course, so the guests left their grapefruit to view the comet through their glasses. The breakfast was resumed half an hour later, and the last guest departed about the time the milk wagons began to rattle through the streets. They all carried away with them small silver telescopes which had been given as favors by Dr. Grace.

There was a comet party on the roof of the Hotel Astor. One party was made up of seventeen guests at the hotel, who had procured binoculars the day before. When Halley's comet first showed itself over the bridge there was a battery of glasses turned to it from this particular roof. At a party on the roof of the Knickerbocker there were many "Oh!"'s when the comet was first sighted. The women all wore their furs, while the men were bundled up and resembled inhabitants of the polar regions. Walter B. Pelton of Springfield, Mass., a guest at the Knickerbocker, watched the comet from there. "The view was well worth waiting for. I was on the roof by 2 o'clock and after I had studied the sky and picked out the spot where the comet was to appear, I amused myself by watching the lights of the city. It was an unusually clear night, and I could see the lights twinkling way out on Long Island. When the comet finally got started I joined with the others on the roof in silent contemplation. Of course the comet looked different to each person. To me the head appeared to be about the size of an electric light globe, such as is used at the hotel, and the tail appeared something like a dim searchlight. I certainly enjoyed the novelty."

Helen C. Mansfield, columnist for *The Bellman,* a magazine for the New York hotel clientele, summarized New York activities of May 18th in her column "The Rounds of Gotham" under the headline, "The Comet in New York—Frivolous Watchers." She wrote caustically and critically, like a social censor:

> New York's basic and ineradicable frivolity came interestingly to the surface last night, apropos of the much-heralded engagement between our globe and the tail of the late Mr. Halley's comet. Of course this occasion was somewhat more than a purely local event, so far as New York was concerned; yet I doubt if in any other city the incident, as we may now safely call it, was observed in quite the spirit that was manifested here; so that the local phase of the matter has an interesting bearing on what Arthur Symons would call the psychology of cities.
>
> What was the situation last evening? The inhabitants of the earth were to experience a cosmic event of a character

which, one might think, would cause even the most superficial to experience a pretty definite sense of awe. Yet what did New York do on "Comet Night"—at least that portion of it which, in the eyes of most of the other inhabitants of this country, is typical of the metropolis? Why, it seized the opportunity to make a kind of New Year's Eve carnival of the occasion. The expensive restaurants were jammed by diners who had engaged their tables days ahead, as they do New Year's Eve; and there was the reckless and public consumption of champagne, the feverish and noisy gaiety, which, in the view of the typical Gothamite, represent the acme of relaxation and pleasure.

It was not that these gentle and civilized souls were obeying the Scriptural injunction to "eat, drink, and be merry" in the face of approaching death, for that kind of gaiety has a certain desperate and splendid heroism which is quite beyond the achievement of our most characteristic New York public. No, they were simply grasping the occasion as an excuse for spending money in the most conspicuous and vulgar manner possible. They were not interested in the comet save as a peg on which to hang a stupid jest; and if they were awed by the event, or in the slightest degree impressed by its circumstances or potentialities, they gave no sign of it. The New Yorker (and I am still, let me make clear, talking of the huge and noisy and conspicuous majority, not the fine and worthy minority) is not happy unless he is sitting in the limelight spending money: this is his idea of amusement, of "a good time." The fact that an encounter between a planet and a mysterious cosmic force should be regarded by him as merely an excuse to dress up, go to the Martin or the Knickerbocker or Stanley's, and "whoop up," is interesting, amusing, and not a little pathetic.

These are strong words, marking the strong sentiments of the Social Register set. The moral tone of Mrs. Mansfield's diatribe sounds excessive and overblown in light of the crime—making an occasion out of a harmless celestial event.

Flatbush?

A "Halley's comet party" became the inevitable fad in the social life of Flatbush. The idea originated with Miss Gertrude T. Cruser of 1904 Glenwood Road. Early on the morning of May 14th a number of Miss Cruser's friends, making a party of thirteen and thus defying superstition, assembled at her home to try to get a sight of the comet. All wore black robes and masks. The parlors were decked out with imitation spiders and their webs, and there was a faint and ghostly light in the rooms and other effects to give a suggestion of weirdness. At the time of the rising of the comet they all adjourned to the rear verandah, where they tried to find it in the skies with telescopes, but weather conditions were not favorable. The "comet party" idea became a fad in Flatbush social circles. Some artful Flatbush men were already scheming to adopt it and shelve the time-worn excuse of an installation of officers at the lodge, offered upon daybreak arrivals at home.

Seattle

"Guests are cautioned to have their house slippers and a warm wrap on when the bellboy rings. Leave the room quietly and join party at the elevator; only two trips made to the roof by the elevator each morning." These and similar instructions startling to the uninitiated were posted on doors of many Seattle hotel rooms during the last few days before May 18th; they were following the lead of the Archibald, which on Sunday night opened its roof to comet gazers. Practically every guest took advantage of the comet call of the bellboy and a ghostly procession, pajamas and kimonos showing prominently as the furtive breezes wafted around the cornices of the Archibald roof. Barring the headache which was sustained by a guest whose neck came in violent contact with a wire, and the inconvenience to which one young lady was put owing to the fact that she forgot her slippers on the graveled roof, the party was pronounced most successful. The Hotel Washington, the Hotel Lincoln, and other hotels were also the site of little comet-gazing parties on Sunday night.

Berlin

Countless "comet picnics" were organized for the big night in Berlin, some on Kreuzberg Hill, an eminence on the southern outskirts of the city, and others in the forest of Grunewald, lying between Berlin and Potsdam. Many thousands of persons congregated in the parks, all the open air restaurants laid in large stores of provisions, and special steamship excursions on the lakes around Berlin, starting at midnight and lasting till 6 a.m., were organized. Germany, indeed, was transformed into one large camp of comet seekers. There were women's hats with mother-of-pearl hatpins shaped like comets, comet walking sticks, umbrellas, textile goods, and brands of spirits for sale.

Geneva

American visitors headed a rush to the highest Alpine resorts in the vicinity of Geneva in the funicular cars to view the comet. Many private comet dances were given at the hotels, starting at midnight. Most of the hotels and restaurants stayed open all night. The Swiss Aero Club arranged a midnight balloon ascent from Lausanne. The balloon carried two Geneva astronomers and their instruments.

Baguio, the Phillipines

In the society pages of the *Manila Times,* amidst notes about dinner parties to honor departing members of the community and ads for hotels in Shanghai and Yokohama and the Provinces of the Phillipines, appeared this brief article:

Baguio Dancers Wait for Comet

Baguio, May 17—Saturday night the weekly ball at Government Center was made a comet dance and dancing was indulged in until about a quarter to three Sunday morning when the dancers waited until the comet's nucleus showed itself over the eastern horizon. The night was beautiful, and the air clear, and the comet was a most beautiful spectacle as its tail cleft the sky in weird brilliancy from the horizon more than half-way to the zenith. The ball was largely attended.

In the outposts of culture and civilization such comet-related events took place. The colonial empires and the commonwealth bastioned in Africa, the Pacific, the Mid-East, and parts of Asia where transplanted Europeans inhabited the frontiers all kept up the social graces of the homeland.

29. HALLEY'S COMET AND THE "THIRTEEN CLUB"

On Friday, May 13th, S. A. Mitchell, adjunct Professor of Astronomy at Columbia University, spent the evening with 169 members of the Thirteen Club trying to pluck the sting from Halley's comet. Thirteen members sat at each of the thirteen tables. At the beginning of the meeting each member spilled salt on the table, and as many as possible sat under big umbrellas raised at the head of each table.

Prof. Mitchell made a contribution to superstitions early in his address on the comet, saying that the comet would rise at 2:47 o'clock on the 13th and that these figures added together make thirteen. He also said this morning would be the most favorable one for seeing it until after it passed the sun and reappeared on the other side about May 21st or 22nd.

Mitchell pointed out that a comet nearly always just precedes or follows a war. Since there are an average of three comets a year, it would be difficult for a comet to miss a war.

Prof. Mitchell said he was sure that all members of the Thirteen Club were looking forward with keen interest to next Wednesday, when the earth would pass through the tail of the comet.

"There is no doubt in the world," he said, "that the comet's tail contains cyanogen gas, found in prussic acid, a poison so powerful that a drop on the tongue would cause instant death. The spectrum shows it plainly. Very probably some of you would like to take to a dugout or subcellar that day. If you are really so afraid that you are all going to die on that day, I hope you will make your bank accounts over to me, as I am willing to take my chances."

30. THE COMET AND CRIME

Who would have thought that the comet's appearance would have had any effect on the incidence of crime? Yet in strange ways it did. There were instances of crime that took place because people were away from home viewing the comet, crimes prevented because people were up at unusual hours to look for the comet, and one instance of confession and restitution brought about by the comet's presence.

Two well-dressed young men passed through Towaco, N.J. on April 29th telling everybody they met that the next morning, between midnight and sunrise, Halley's comet would be visible from the top of Waukaw Mountain. They said the scientific school which they represented would give prizes for the best amateur descriptions of the comet and its tail.

There was not much interest in these prizes until the word ran around that Lilly Lautergan, Cyrus Lautergan's more than talented daughter, was going to the top of the mountain, with her easel and palette, and there make a sketch of the comet, its tail, and all its environment and background.

Mary Vanderlip, daughter of Josh Vanderlip, said that the teacher at school had declared that, so far as sky coloring was concerned, she was equal to and most likely superior to Lilly Lautergan. On the morning of April 30th, Waukaw Mountain top was covered with Towaco's beauty and chivalry and chaperonage. Miss Lilly and Miss Mary were ready with their palettes, with bets on the winner.

A good time was *not* had on Waukaw Mountain top, Halley's comet was *not* seen, and when Towaco's citizens got back down to their homes they found that their chicken coops had been looted. Cyrus Doolittle was missing 300 fowls; both the Lautergan and Vanderlip families were heavy losers.

"Halley's comet can go to the dickens, as far as I'm concerned," said Mr. Lautergan after he had counted his losses. "Doggone comets, anyhow."

"Don't be rude, father," said Lilly. "It hurts me. Those young gentlemen were not the chicken thieves."

By the same token the comet can be credited with preventing crimes. A. M. McKnight of the Bayside section of Queens was looking for the comet at 3 a.m.

from a window when a movement on the roof of his brother's home attracted his attention, and he saw that someone was trying to break in to the house through the roof. Looking no further for the comet, he ran to the telephone and called up his brother, E. Scott McKnight, to warn him. Still in his pajamas, E. S. McKnight grabbed his large calibre revolver and hurried to the upper floor. From the window he looked at the roof. But the burglar had heard the telephone ring, and had immediately run off. McKnight heard the man scrambling down from the porch and fired at him. Then he chased the man, who got away.

In Denver, a "peeping Tom" was captured shortly after midnight on May 18th at the home of the Flavin family where several friends had gathered to watch for Halley's comet from darkened windows. For several nights the residents in the vicinity had been annoyed by a "peeping Tom" who grew bolder each night. Three times he had been chased but escaped. On Tuesday night, stealing along the wall of the house, peering into each window as he walked, the head and shoulders of a man attracted the notice of the watchers. Miss Florence Flavin, eighteen years old, whispered that the man must be the "peeping Tom" whom she feared. Two men left the house by different doors. Mr. Flavin leaped on the man whom he found crouching in the shadows. The police were called and the man was turned over to them when they arrived.

The comet was also responsible for instances of confession and restitution. Frank L. Hayes, living in Denver, traveled to Greeley, Colorado to pay debts of twenty years' standing in order to put himself right for the Judgment Day. Having left Greeley in 1890 owing debts, Hayes candidly told his creditors the reason for his action. Believing that the coming of the comet presaged the end of the world, he was anxious to have a clean record when the end came. Dr. R. F. Graham, the Henderson & Kendel Grocery Company, a druggist and several old-time merchants all profited by the Denverite's contrition. Dr. Graham remembered the man as soon as he saw him. One of Mr. Hayes' creditors, a Mr. Henderson, had died many years ago and his son had succeeded to the business. Young Henderson could not find the account on his books and refused to take Hayes' money. The man not only begged Henderson to take it, but insisted that he look over his father's old papers. The books containing the entry had been stored in a cellar for a decade. After a two-hour search Henderson found the bill for $75.00, the amount Hayes had wanted to give him in the first place. All together, Hayes paid out about $200.00 to his creditors, who appreciated the practical form his contrition took.

31. HALLEY'S HALL OF ODDITIES

This chapter marks those incidents just plain curious and having a twist of the comical or humorous about them—little firecrackers among the bombs.

Halley's Best

The *Chicago Tribune* reported a special meeting of the Chicago General Committee for the Reception of Halley's comet. Professor Graham Taylor read a report from Oxford Professor Turner, stating that if people wished to bottle some of the air on the night on May 18th they could hand a part of the comet down to their grandchildren. On behalf of the Committee, the Treasurer was directed to buy fifty dozen bottles of champagne for May 18th. After the champagne was finished the bottles would be filled with Halley's best and recorked to be taken as souvenirs.

Letters Accuse Astronomers of Lying

Harvard Observatory was flooded with letters about Halley's comet. Most of the writers did not know what a comet was. One writer asked how much a comet cost, since he was interested in buying it. Some people wanted to know how best to avoid the comet and if there was any cure for it. Some wrote to say they were going to recommend to Col. Roosevelt that the comet be shot on sight.

"Really, you know," said Professor E. C. Pickering, "they seem to think we are responsible for it. Insinuations in several letters we have received made us fear we were being denounced as undesirable citizens.

"The advocates of a flat earth have risen again in their might and have informed us that there is no comet because the earth is flat. It is so. So it is. There is no room for argument. But the advocates of a flat earth we have always with us, while the comet has brought us into communication with a score of new writers, who prefer to get at us this way: 'You know the comet is going to hit the earth and end the world. Why don't you tell the truth about it?' they write.

"Our motive in withholding information is frankly sinister. We want to influence the stock market, or have been bribed."

A Bet on the Comet Hitting the Earth

Mr. Bramwell Davis, who was called the Astor of Great Neck, met former New York City Police Captain James M. Churchill on the train coming to New York. They talked about the chances of the comet colliding with the earth. Davis said, "Churchill, I'll bet you $10.00 to $5.00 that she won't hit us."

"You're on," answered Churchill. "That's a bad bet for you if you win."

"Why? The odds?"

"Nothing of the kind. How will you collect if you win?" asked Churchill.

90 Year Old Man Marries for First Time: Halley's Comet in Sight

George Durkin, ninety years old, of Guilford, Connecticut picked Friday, May 13, to marry a woman who broke the United States record for the quickest divorce. Durkin saw Halley's comet the last time and wanted to be married while the comet was visible.

The lady in the case, Durkin's housekeeper, a Mrs. Grisvold, was divorced several months before. Mrs. Grisvold was half Durkin's age. Mr. Durkin was a civil war veteran and quite wealthy. He told a reporter that his marriage was the result of a comet courtship.

Comet Pills

The comet brought a small fortune to an old medicine man in Port-au-Prince, Haiti, according to the story told by passengers on the Atlas liner, *Alleghany.* He dispensed a charm more potent than the rabbit's foot among the Black longshoremen, farmers, and servants. Those who had any money hurried to the voodoo man to buy his anti-comet pills which were guaranteed to protect the purchaser from all cometary evils. The price was $1.00 a box.

The comet scare was widespread and the sale of anti-comet pills enormous. Those who bought them immediately loaded their systems with the pills and then walked with confidence among their frightened and unprotected neighbors. The old Negro did a land-office business and could hardly keep up his supply of sugar pellets.

The pills had the effect of quieting the fears of the ignorant who believed a pill or two under their belts would allow them to walk unscathed over the ruin of the world.

Calves' Markings Due to Halley's Comet

Friday the 13th was a lucky day on the farm of Amos Rhoades of Pitcher, N.Y. On that morning one of his prize Dorsetshire cows gave birth to four well-formed calves. Usually cows bear only one calf a year. Two of the calves had unusual star-shaped markings on their foreheads. Farmer Rhoades' dairy maid said the markings were due to Halley's comet.

Took Down Lightning Rods

Somehow the belief spread in Neemah, Wisconsin that lightning rods would attract dangerous substances from the comet's tail. Many farmers in the vicinity removed the lightning rods from their barns and homes as a precaution on May 18th.

Boy Awaits Comet

Tommy Hotchkiss, twelve years old, of Taylortown, New Jersey returned to his home on May 14th after being away twenty-two hours. During that time search parties had scoured the woods and mountains in the neighborhood. Luke McElroy, a janitor at the public school, discovered the boy fast asleep in the school tower when he went there to ring the school bell to give a general alarm for the boy. Tommy had armed himself with four sandwiches and a dime novel when he went to school on Friday. When classes were dismissed, instead of leaving with the other pupils he slipped up to the school belfry and waited for the comet to come. He became greatly frightened after it got dark. The owls hooted and the wind whistled through the belfry, but he was afraid to move from where he was. He stayed awake as long as he could, and then fell asleep.

Comet Reconciles Estranged Couple

On May 7th Mrs. Gertrude Sanders of 84 Brown Place, Jersey City, argued with her husband James over compliments he had paid her sister the evening before at a social gathering. Mr. Sanders allowed his wife to grow jealous and quarreled as a joke, not realizing that she had taken it seriously. Mrs. Sanders left home after breakfast leaving a note for him telling him not to worry, she could earn her own living. Mr. Sanders offered a thousand dollar reward for information leading to her recovery.

On the night of May 18th Mrs. Sanders telephoned her husband from New York, asking to meet her at a boarding house on West Thirty-fifth Street where she had been staying since she left home. He went there and found her in a nervous state through worry, but eager to be reconciled with him on her twenty-second birthday. People had told her that day might be the last day she or anybody else would have on earth.

VII
The Voice of the People

32. THE NAKED EYE: REMINISCENCES AND REFLECTIONS

R. BUCKMINSTER FULLER
(Born July 12, 1895, Milton, Massachusetts - Died July 1, 1983)

I was fifteen at the time of Halley's comet's appearance. I lived in Milton, Massachusetts and attended Milton Academy. It was an extraordinary school. At that time it was exclusively a Harvard preparatory school. One member in my class went to Yale and broke the long-standing tradition. It was at the highest rank scholastically of all the Harvard preparatory schools. So things like Halley's comet were of great interest and were discussed in the very best scientific way. We were all excited.

My father had a beautiful camera for those days, a Kodak Bull's Eye. It took 4″ x 5″ film. There were not really many good cameras like that in those days.

The comet was visible for quite a few evenings. I'm sure I didn't take the photograph the first evening. I remember coming out from the house to take the picture quite early in the evening. We had quite a wooded place at our house. The comet was showing between a big pine above some oaks. I was able to get a good view of it. I got a very good picture.

Nobody in my community was superstitious about the comet. We knew it was coming and there it was. It was just a normal phenomenon, like everything else.

* * *

When I was twenty-eight Hubble discovered another galaxy other than our Milky Way. Up to six months ago we had discovered two billion more galaxies and within the last six months* we had discovered two-hundred billion more galaxies with the radio telescope.

I have seen remarkable changes in my lifetime.

**Conversation held on April 3, 1983.*

Cecil Lucas
(Born Elmira, New York, July 6, 1900)

I saw Halley's comet with the naked eye. There weren't many instruments around at that time. A few field glasses that came from the Civil War and a few opera glasses, but they were very low power.

The comet itself was scary. My recollection of it is that it passed over very fast, much faster than stars do. It was there and it was gone very soon. There were several appearances of it at that time. On some nights you wouldn't be able to see it on account of cloud conditions. The head was awesome. It looked like a very large star. The tail seemed like it would swish back and forth.

My recollection is most people took it pretty easy, they didn't get too excited. Of course there's this certain fanatic group that got on the bridges down on their knees praying and all this sort of thing. Some people figured if that tail ever hit the earth it would destroy the earth completely.

Church groups were preaching the end of the world. They were preaching on every street corner, "Get your soul saved now! This will be your last chance!" I was handed little tracts by the street preachers to take home and read. My folks didn't believe in those things so I didn't either. It was scary.

The next thing they said that if you were on water the chances are you might be able to survive. So they encouraged everybody to get out on the bridges. Well now in Elmira there's a bridge for every street. There's probably twenty bridges within the city limits running across the river. The Chimong River goes right through the center of town. They got so many people out on the bridges that the bridges started to collapse. The bridges were weakening.

At that time my father was the superintendent of the American Bridge Works. He had to get the people off the bridges again.

There was a proclamation in the papers telling people not to congregate on the bridges. After that there was no risk, no trouble with the bridges.

After the comet finally disappeared, why there wasn't much talk about it. These religious people were upset that the comet didn't destroy the earth. In other words their forecasts were disapproved. Of course they didn't like that. So they started preaching that it would be the year 2000 before the comet would appear again. They said this one was a warning and the next one would actually strike the earth and destroy it then. That's what they were preaching after their forecast fell through.

After the comet disappeared people forgot about it.

<h3 style="text-align:center">Froika (Frank) Fershter,
(Born Bonderova, Ukraine, April 24, 1896)</h3>

It was a Sunday in 1910. An automobile passed by. It was the first time the people had ever seen an automobile. I had already seen one because I had been to different places.

The peasants were smart people, they were intelligent people, but they weren't educated. They didn't know what to make of it. It was the time of Halley's comet. Some said it must be a part of the comet. Someone said it must be something that brings the end of the world and someone said a devil sits in there and operates the machine.

I don't know how the peasants knew about Halley's comet. The town where I lived had about fifteen blocks and I don't think amongst all of those people there were ten people that could read or write, but everybody knew that something was coming; they just didn't know what it was.

When they spoke in our town about the comet everyone thought, "Who knows what will happen?" But after I read about it a couple of times I knew that it just comes naturally. The comet comes every so often.

I had a book that described Halley's comet. Somehow the peasants knew that I had it. So they came up to me and said, "Froika, we want the book that tells about the comet that will come and destroy the earth." So I told them it's not as bad as that. It won't destroy anything. I couldn't tell them that nature provides the comet. If I told them that, I'd have to explain all these things and I didn't know anything about them myself. So I told them it's God's doing and that's all. He'll see that we're safe. What can you tell an illiterate person?

I remember just like today. There was a bunch of Russian boys and girls visiting our house. That evening we were sitting around the table and drinking tea and talking. One of the fellows said goodnight. He was going home. All of a sudden he comes back and he knocked at the window. He said, "Come on out, ladies and gentleman and you'll see something."

We went outside and looked around. The comet was there. It was almost as big as the moon. The comet and its tail covered up almost the whole horizon. It made the night like a street now looks when the street lights are on. It was beautiful.

I still remember as I walked out of the house it looked to me like the fiery tail was all on me. I wasn't scared, I was stupefied. I didn't think of God, I didn't think of angels, I didn't even think of nature. I thought just of that thing. It looked like gold. The comet didn't have any certain quality. It just looked like something the imagination would make. It was a dream. It was like being in space. It was such an experience, I'll never forget it.

33. HALLEY'S COMET
AND
THE CHRISTIAN LITERATURE SOCIETY IN CHINA

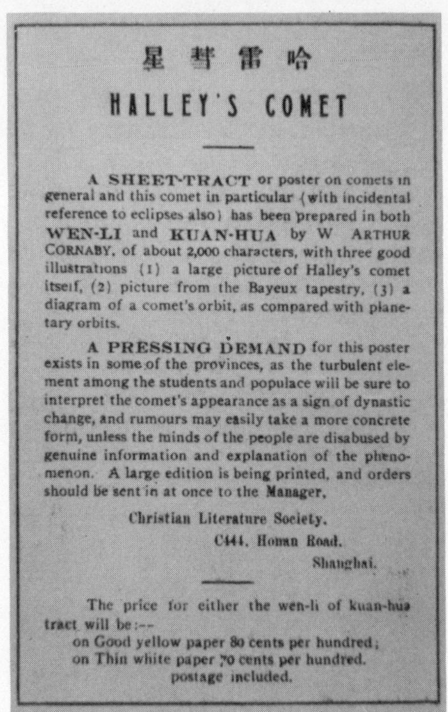

哈雷彗星

HALLEY'S COMET

A SHEET-TRACT or poster on comets in general and this comet in particular (with incidental reference to eclipses also) has been prepared in both WEN-LI and KUAN-HUA by W ARTHUR CORNABY, of about 2,000 characters, with three good illustrations (1) a large picture of Halley's comet itself, (2) picture from the Bayeux tapestry, (3) a diagram of a comet's orbit, as compared with planetary orbits.

A PRESSING DEMAND for this poster exists in some of the provinces, as the turbulent element among the students and populace will be sure to interpret the comet's appearance as a sign of dynastic change, and rumours may easily take a more concrete form, unless the minds of the people are disabused by genuine information and explanation of the phenomenon. A large edition is being printed, and orders should be sent in at once to the Manager,

Christian Literature Society,
C144, Honan Road,
Shanghai.

The price for either the wen-li of kuan-hua tract will be:—
on Good yellow paper 80 cents per hundred;
on Thin white paper 70 cents per hundred.
postage included.

from
Collier's Magazine

In November of 1909 Donald McGillivray of the Christian Literature Society visited some cities in China's far interior. During an address he gave to officials

110

and students he mentioned that Halley's comet would soon be visible. Mr. McGillivray recounted the audience's response, his experiences, and what action it led him to take. Wrote Mr. McGillivray:

> The address was, much to my surprise, interrupted at this point by excited inquiries for further particulars. This experience gave birth to the idea of a wide comet propaganda, and on my return to Shanghai the idea was set afloat by the preparation of a striking poster by the Christian Literature Society. From September, 1909 onward there was growing unrest throughout the Empire, though not as a result of the comet. The unrest pointed up the need of removing this element of additional danger, and lent wings to the propaganda of enlightenment. A spark, so to speak, from the comet, finding the air already electrified, might well precipitate widespread disaster.

McGillivray sent letters to six Tract Societies, pointing out the importance of preparing the minds of the Chinese people for the comet. If this work would not be undertaken, secret societies would use the comet's appearance as an opportunity to cause unrest and riots throughout the country. According to the Chinese,

> The presence of so many foreigners scattered up and down through every province of China rendered the danger of riots all the more likely, especially if designing revolutionaries point to the comet as a signal of probable successful rebellion.

Additional tracts were prepared at Hankow, Shanghai, and West China by other Tract Societies. All the pamphlets were sold very cheaply and long before the comet became visible to the naked eye. Chinese newspapers reprinted the tracts along with editorial support for the views they expressed. Copies were sent to all high officials of the Empire, and governors and inspectors of education from Yunnan to Manchuria ordered 33,000 copies to distribute to their subordinates. Merchants bought quantities to give their customers.

The Christian Literature Society posted tracts at 3,500 telegraph and post offices throughout the Empire of China. Copies were posted in many schools. Two thousand were sent to Korea. Altogether 115,000 Christian Literature Society tracts were sold: 65,000 of the Chinese Tract Society and 22,000 by a Dr. Hallock were posted all over Shanghai. The missionaries of Kansu produced a special poster of their own, which they reported did much to quiet the people.

As Mr. McGillivray predicted, when the comet appeared inflammatory circulars appeared on walls and fences all over the Empire saying that the comet was a sign that the foreign powers were going to carve China up like a melon and divide it among themselves. The posters, prepared by secret revolutionary societies, called upon all students and other loyal Chinese to help avert this calamity. They were exhorted to enroll as volunteer soldiers, learn the use of arms, and prepare to fight the foreigners and drive them from China. But the Christian Literature Tract Society's pamphlet prepared the people for the comet. The knowledge that the earth was to pass through its tail on May 19th had less effect upon the Chinese than it did upon people in Russia and even the western United States.

Mr. McGillivray concluded his report with observations which remind us that his object was to convert the Chinese to Christianity. He claims that by predicting the comet's appearence well in advance and with great accuracy,

> Astrology, geomancy, and horoscopy have been sensibly weakened. For this every foreigner, syndicate, company and merchant should be forever grateful to the missionary body, and this because these superstitions are their most formidable enemies in China. To multitudes of Chinese have come their first ideas of the solar system and its great Creator, and new and striking attention has been called to Christianity in China.

34. HALLEY'S COMET: CAUSE OF BAD WEATHER?

> *Staggering questions, not unlike some of the posers with which children stump their elders, have been put to many scientific men in relation to the supposed effect of Halley's comet upon our pleasant planet. Over and over again have the wise ones been asked, "Was the comet responsible for the unprecedented atmospheric and seismic disturbances that brought so much woe to the human race during the first half of the present year?" Is a man a superstitous fool to blame any of those cataclysmic upheavals, devastating inundations, terrible typhoons, frightful and unseasonable frosts and snows, or unprecedented heat, upon the comet? On this misty subject of cometary influence upon mundane affairs are the scientific any wiser than the unscientific?*

> ("In the Year of the Comet" by
> Bailey Millard, *The Technical
> World Magazine,* July, 1910)

Henri Deslandres, Astronomer

In the second week of May, the French astronomer, Henri Deslandres of the Meudon Observatory, delivered a talk to the French Academy of Science of which he was a member. The French astrologer and prophetess, Madame de Thebes, had predicted in January the disastrous spring floods of 1910 in France, Italy, and Germany and the coming of Innes' unexpected comet. Commenting on Madame de Thebes' predictions and her connecting Halley's comet with the floods, Professor Deslandres said:

> However distant comets may be, it is not at all impossible that their enormous tails, measuring 75,000,000 to 125,000,000 miles in length, may come in contact with our atmosphere. The theory that a comet may disturb the atmosphere of the earth, causing rains of great duration, and consequently inundations and the sudden overflow of rivers, is not at all absurd. It can be sustained by scientific reasoning.

Alas, scientific *reasoning,* by which Professor Deslandres must mean logic, is very different from *science.* Indeed many pamphlets, based upon scientific *knowledge,* were published demonstrating the impossibility of any significant effects being experienced by the earth.

Irl Hicks, Meteorologist, Minister, and Publisher of *Word and Works*

Since 1865, *Word and Works* forecast the weather and preached the revelation that God's good and gracious power is manifested in the position of the planets, in weather, earthquakes, and volcanos. In addition to attributing activity in the earth's crust to God's proving His presence and power to us, *Word and Works Almanac and Magazine* also forecast the weather. Almanacs have a venerable history of long-range weather forecasting, *Hicks' Almanac a Year in Advance* being among the best. A proficient forecaster, Irl Hicks was also a minister and prophet who skillfully preached his message out of a cloth weaving meteorology, astrology, astronomy, seismology, and religion. Naturally, Irl Hicks had something to say about Halley's comet and the weather, and he said it in an essay entitled "The Comet and the Crisis." Following an accurate, factual account of the comet's path, Hicks claimed:

> Whatever effect (the comet) may have on the Earth, will come by way of the Sun, and phenomenally warm weather through March, the convulsions on the Sun, with volcanic violence, earthquakes and abnormal auroral lights, beyond question are largely the results of Halley's Comet in its perihelion flight around the suburbs of the Sun. Whatever other results may follow from this close visit of the great celestial wonder, we do not hesitate to declare, *will not be accidents. The Almighty hand that created and upholds the universe is guiding the comet.* (Italics mine.)

Hicks then takes on the dress of a prophet; with a single swipe of the pen he excoriates the materialism of the day:

> Whatever men may say, we firmly believe that God intends to arrest the attention and thoughts of men by sending such marvelous wonders alongside our little world, and in order to do it, if need be, He will cause the comet to blight the Sun and shake the Earth. When vain man flatters himself that he has succeeded in eliminating God from the universe—when the stayed order of nature has crystallized the minds of men into *scientific* unbelief and blindness, God can and *does* call forth comets, and whatever other wonderful and alarming agencies He pleases out of the deep, silent, eternal heavens.

Refuting these very "materialists," he distinguishes between "superstition" and true religion.

> It is *not* the work of "superstition" to look for the hand of God in the comet, but it is the part of holy reverence and faith to

learn the lesson He would teach—the stupendous fact that a *Divinity* presides over all things in the Earth and heavens. With such reverence, such spiritual discernment, such humble, loving faith and confidence in the good and gracious God, His people can with serenity and calmness look upon any exhibition of His power, or enter into the mysteries of any shadow He may throw upon the world.

Mr. Hicks' views were widely read and his opinions were widely held.

Military Aide to the President

Archibald Willingham Butt, Military Aide to Presidents Taft and Roosevelt—and later a passenger on the ill-fated *Titanic*—wrote wonderful private letters to his family which were published eighteen years after his drowning. Butt's letter of May 16th, treated three subjects: playing golf with President Taft the day before, the cowboy outfit he was going to wear that night to the Preston Gibsons' Indian Party, and Halley's comet's effect upon the weather. About the weather he wrote:

> The entire spring has been so chilly continuously that many think that the comet is drawing the heat rays of the sun away from the earth. Certainly we have no record of such a cold season before. Before the comet appeared last March we had extremely hot weather, and it was thought that there would be a general exodus from the city in May. Most of the houses have relighted their furnace fires. They have done so at the White House, which I think is a mistake, for every room has an open fireplace in it, and the wood logs are much more healthful at this season of the year than furnace heat.

The People

In Chicago on April 12th sudden darkness, descending for two hours, created terror among the more ignorant, who attributed the phenomenon to Halley's comet. In street cars women became hysterical, and in the foreign quarters policemen put the fears of the nervous to rest. The Weather Department reports from different parts of the city explained that the darkness was caused by a combination of rain, wind, and smoke.

At sunset on May 18th a sudden boom of thunder followed immediately by the darkening of the sky and a heavy downpour of rain gave the nervous folk of Boston, who were expecting strange happenings from the comet, a mild shock, but the scare quickly ended. The sky had been obscured by clouds all day. After the shower it was clear, but the bright moonlight prevented any observation of the comet's tail.

Did the comet cause bad weather? The learned professors said no; journalists and writers said "maybe" or "certainly" and dramatized scenes of human suffering caused by the comet through unseasonable weather—hot when it should be cold, drought, floods—and volcanic eruptions. In many places the people acted as if the comet had caused the bad weather. No doubt we will take another look at the question as the comet draws closer. In our own age we point to acts of man as the cause of changes in nature. And, today, meteorology shows us the weather globally from the eye of a satellite.

35. MISTAKEN FOR THE COMET

During the days when the comet was present, freak occurrences and accidents were sometimes attributed to the comet by the fear-stricken.

Comet Scare in a Trolley Car

On the night of May 16th, passengers in a crowded north-bound Eighth Avenue trolley car in New York were all convinced that the tail of Halley's comet had side-swiped the earth. As the car was passing through Columbus Circle, the roof broke with a crash and every pane of glass shattered.

"It's the comet!" exclaimed a woman passenger. The same idea possessed every other passenger, and there was a rush for the doors.

What actually happened was that employees of a motor car company were testing a new cooling fan for automobiles when one of the fan blades got loose and flew through an open fifth-floor window. The blade, weighing about twenty pounds, was turning at 1,000 revolutions per minute when it broke off. The piece of metal went up at least 600 feet before starting downward.

The large hub end of the blade stuck in the roof of the car, the blade piercing through. The shock broke every window of the car. The glass flew in all directions, but miraculously no one was hurt.

Two Explosions Thought to Be Caused by Halley's Comet

A monstrous explosion of a large quantity of nitroglycerin stored in a magazine fourteen miles from Pittsburgh took place on May 10th. It caused McDonald residents to run from their homes, some shouting, "Halley's comet has struck the earth!" At Greensburg, eighteen miles from the scene of the Burgettstown explosion, houses rocked. On the North Side Hills of Pittsburgh a man flying carrier pigeons said his birds were stunned by the shock wave of the explosion. Oil well shooter Frank McCullough was killed by the explosion. He was at the site of the magazine. Not even a fragment of his body was found. McCullough's team of horses was also blown up.

In Paris, the Bicetre Hospital was the scene of a terrific explosion on May 16th. An infirmary keeper, haunted with the idea that he was an inventor, was experimenting with nitroglycerin when it exploded. Investigators discovered enough explosives in his basement to blow up all of Paris, but they did not go off when the nitroglycerin exploded. The hospital's inmates, all elderly, became so imbued with the idea that Halley's comet had struck the earth that they rushed for the gates.

Practical Jokes and Jokesters

May 18th, 1910 brought forth its share of practical jokers, several creating fear and panic; one was utterly benign.

Between eight and nine o'clock hundreds of people gathered at the corner of Mulberry and Broome Streets in New York City, gazing at the heavens. Suddenly, someone saw a large bright light descending and shouted a warning. A serious panic threatened. The police reserves were called out from the Mulberry Street Station to calm the crowd and clear the streets. The frightened men and women were driven into their houses, shouting while they watched the descent of the strange light, which was a toy balloon with firecrackers tied on it.

Knowing that many people were expecting strange events resulting from the earth passing through the comet's tail, H. C. Boehn of Roselle, New Jersey staged his own practical joke. He assembled a balloon, some sodium, a time fuse, and a stick of dynamite. Launched from a large vacant field south of the town, the apparatus worked perfectly, rising to 4,000 feet and exploding with a terrifying roar. The explosion was heard for miles. The dynamite ignited the sodium which fell blazing to earth. Midnight comet watchers were thrown into a state of terror. Pandemonium resulted, and it was many hours before the fear of some of the watchers was allayed.

One man made a comet's tail where there was none. Evening clouds hid the comet from view for several evenings prior to May 22nd in the New York and New Jersey areas. So, a searchlight manufacturer in Plainfield, New Jersey, decided to satisfy the curiosity of comet gazers. He mounted a thirty-six-inch electric searchlight, the type used on battleships to illuminate targets at night, on his factory's roof. In the misty sky of the evening the tail of the fake comet shooting upward was an impressive sight.

Ordinary Occurrences

With attention turning to the heavens, there were many, unaccustomed to what their eyes beheld, who mistook ordinary occurrences for Halley's comet. The lights of off-shore buoys or searchlights on tall buildings or planets were taken by some to be the comet. This letter in *The New York Times* is typical:

> To the Editor of *The New York Times*
> Your paper has kept its readers well informed of the movements of Halley's comet. I think this is the first report you have received of its having been sighted in the heart of the Catskills. A party of us at the Silver Stream Cottage arose at 3 o'clock Sunday morning to see the newcomer. At just 3:22 o'clock the fiery ball was seen to make its appearance above the horizon, over the ridge of hills just east of Tonshi Mountain, beneath and just a little to the right of Aquarius.
>
> There was no tail in sight, not even by the aid of powerful field glasses, but at times there seemed to be flashes of brilliancy more intense. We observed the wanderer until 4:15 and during this time it did not seem to change its aspect.
>
> Frederick J. Rehn

The brilliant, tailless object our friend saw beneath Aquarius was Venus, not the comet.

Editor, Times

Not only objects, but smells and effects upon physical states were mistaken for the comet or thought to be caused by it.

Thursday morning, May 19th, James Stuart, chief clerk of the Board of Public Health and Safety for Denver, went into the dining room where he noticed a peculiar smell. He went to the window, opened it, and the smell increased. He told his daughter that the smell came from gases from the comet's tail. Stuart met a couple of neighbors and asked if they had noticed any smells. They had. Stuart imparted his information. The news spread all down the block and Stuart was regarded respectfully in the enlightened neighborhood. All had smelled a disagreeable odor. Friday morning the postman came along, a broad grin on his face. He had heard of Stuart's sagacity all down the block the day before and had made inquiries. He rang the Stuart doorbell.

"Did you smell the comet's tail?" he asked politely.

"Oh, yes, Papa discovered it," said daughter Stuart.

"Um! Ah! Two dogs up in the next block chased a polecat all around this neighborhood Thursday morning," he said. "The dogs have gone to the country."

A similar account appeared in an *Astronomical Society of Wales* column called "Cambrian Natural Observer." Under the headline "The Comet's Odor" appeared two letters from Mr. J. A. Kidd of Cardiff, Wales, in which he reported the strong smell of marsh gas.

> From 2:45 a.m. until nearly 5 o'clock the smell was almost unbearable, and reminded me of the smell of burning peat, only much more pungent. During that time I went repeatedly out in the open air and noted nothing abnormal, not the slightest sign of smoke or anything that could account for the strong smell. I felt very strongly at the time it was due to the gas or gases in the tail of the comet.

Writing two days later Mr. Kidd said:

> Since the notice in the *Western Mail* re smell of comet I have been instructed by people who state they distinctly noticed the strong smell mentioned. One describes it like old boots burning, several like patent fuel, and others like sulphur. One—a lady who sleeps with the windows open—states the smell was suffocating.

In spite of widespread scientific information available to the public, the imagination was forever willing to see the comet as the cause of explosions, lights, and foul smells. This leaning of the mind led practical jokesters to play their pranks. Indeed, there was a certain inevitability about the mistakes people made; highly wrought emotions, particularly fear, coupled with unexpected noises and flashes pointed to the comet as the cause. Blamed as the culprit in the deaths of kings, fancifully pictured in the "half-penny press" as the destruction of the world, it is no wonder that Halley's comet was seen in the pops and fizzles of the night of May 18th.

36. SHIPS AT SEA; BOATS NEAR SHORE

Arriving in New York from Bremen, Germany some of the passengers on the *Kaiser Wilhelm Der Grosse* blamed the comet for the heavy storm they encountered. One of the bedroom stewards said that a merchant had given him only a quarter for looking after his cabin on the voyage, and the bandmaster had found a bad half dollar in the collection, which he put down to the same evil influence. The chief steward said that Schnitzel, the pet cabin cat, had seemed dismal and unhappy all the way across. The cat refused to lap her milk, took frequent trips up the main rigging at night, and gazed earnestly into the darkness, while uttering plaintive meows in German. When asked if he believed this melancholia on the part of Schnitzel was also due to the comet, the chief steward replied that it was either that or remorse for eating the bo'sun's whistling canary on the last passage home.

The comet was still to be seen when the *Kaiser Wilhelm* departed for Europe. Before the ship sailed a reporter sought the opinions of two celebrities making the voyage. The operatic tenor, Alessandre Bonci, said, "I am afraid of Halley's comet and am not ashamed to own it. Of course the scientists say there is no danger, but what do they know about the mysterious laws of nature? I think something will happen, I know not what, but I shall not sleep easy in my berth on this ship until the danger is over."

The actress, Olga Nethersole, said that she was not alarmed for herself about what the comet might do to the ship on the high seas, but she was nervous on account of her pet Mexican hairless dog, Chiquita. "You see, if the tail of the comet touched my poor pet she has not hair to protect her," said Miss Nethersole.

The comet was the one absorbing topic of conversation in steerage and cabin on the Fabre liner, *Germania,* during the voyage to America. The *Germania* came from Marseilles, Naples, and Palermo with ten cabin and 370 steerage passengers. Captain Boulak reports that the comet was visible every clear night of the voyage. During the first part of the voyage the steerage passengers, most of whom were Italian, were disturbed about the comet. On the night when it was seen at its best they were greatly excited. Crucifixes, rosaries, and Bibles were in many people's hands, and groups could be seen on deck praying loudly for deliverance from the evil which was bound to follow the coming of the celestial visitor. Word got about among them that the great danger was to come when the tail of the comet hit the earth. Prayers redoubled. All the third-class passengers wanted to reach land before the catastrophe. They rejoiced when the *Germania* came up the bay to New York City.

In 1910 people all over the world crossed rivers and bays on ferries. How did the comet appear from their vantage point?

Captain Stephen Tinsler of the Jersey Central Railroad ferryboat *Plainfield* spied the comet rising above the Singer tower as the boat was lying in her slip at Communipaw before starting on her 3:15 o'clock trip to this city. Pilot Smith carried the word to the main deck and the market gardeners, teamsters, and a few belated train travelers all crowded to the forward gates for the best view. In the pre-dawn hours the passengers saw Halley's comet poised above the city in the eastern sky as they looked up from the harbor in the lower bay. It was at its brightest soon after three o'clock. Ferrymen and their regular owl travelers, who had watched for it for weeks, found it larger than they had expected, but less brilliant. The tail, sweeping away to the southeast, could be plainly traced from three to four times the apparent breadth between the horns of the fast-waning moon, which hung below, and to the south about on a level with Venus, the comet's closest neighbor. Mackerel clouds veiled the display soon before four o'clock but until nearly a quarter past or until the dawn was well advanced, the pale outline of the visitor could be traced, like a streamer from the Milky Way.

37. HALLEY'S COMET AND THE HUMAN EYE

Ordinarily events which capture the world's attention—floods, earthquakes, shipwrecks, wars—give humanity opportunities to act "heroically" as if this were a valuable measure of our species' potential. Rare among events to capture the world's imagination, Halley's comet's 1910 appearance contributed to a vision of the beneficence, harmony, and beauty of the universe, of which the earth is an insignificant mote.

One little New York town hung a canvas sign above the main street saying "Welcome, Halley's Comet." Everywhere there were lectures on the comet and astronomy for every kind of audience. Preachers preached on it, editorialists editorialized, illustrators made it their subject, poets fashioned verse and "philosophy" around it. But the genuine thrill, of course, was in actually *seeing* the comet. And on the night of May 18, when the earth was to pass through the comet's tail, the greatest number of people *ever to share in a single event watched with mingled fear and awe.* The eye was the sole organ through which humanity experienced the comet. What was actually seen? In "The Comet as a 'Sign in the Heavens'" was described what watchers of the skies would see:

> The moon will be nearly full [on May 18] so that its light will obscure the comet, which, however, should become easily visible with the naked eye in the early evening sky, shortly after sunset, commencing with about May 20. It will then be moving rapidly back toward the east, getting nightly farther from the sun and staying longer visible, until, as it grows gradually smaller, it fades away in the distance, not to be seen again by mortal eyes until 1985.

No sound, smell, or texture—the comet was perceptible only through the eye. Twenty-six years after the comet's appearance Mary Proctor wrote, "[Halley's comet] left a lasting impression on the human eye, adding renewed interest to its long life history of more than two thousand years." *A lasting impression.* Everyone who saw the comet never forgot its majesty, its apparent power and energy, its palpable mystery, its nearly inexpressible loveliness. By the highest and lowest these impressions were cast in words. The eminent astronomer Dr. Barnard, who admitted Miss Proctor to Yerkes Observatory on that night, now generations ago, when she first glimpsed Halley's comet commented:

> It would have been a great satisfaction to know that everyone who saw this wonderful object, did so with the same feeling of elation and wonder—one would almost say veneration—with which the average astronomer regarded this beautiful and mysterious object stretching its wonderful stream of light across the sky.

And so this telling of the 1910 apparition closes as we await the comet's arrival, await "this beautiful and mysterious object," the comet called Halley.

HARPER'S
WEEKLY
EDITED BY GEORGE HARVEY
SEE
HALLEY'S
COMET
May 21 1910   HARPER & BROTHERS, N. Y.   Price 10 Cents

SOURCES

Chapter 1. Who Was Halley and Why Is a Comet Named for Him?
The brief description of Halley's life and work comes largely from Colin A. Ronan's *Edmond Halley: Genius in Eclipse.*

Chapter 2. "Halley's Comet and Its Great Journey"
The school children of America read the description of the comet and its orbit in *St. Nicholas Magazine* for May, 1910. The data the chapter reports were available in many magazines and newspapers. Sources for sky maps are provided in the text.

Chapter 3. "The Comet's Poisonous Tail: The 'Terror Occasioned by the Near Approach of Halley's Comet'"
The New York Times was the source for the reprinted article, the editorial, and the remarks of astronomers regarding the cyanogen in the comet's tail.

Chapter 4. "1910: The Year of Celestial Fireworks"
The New York Times reported the sighting of comets 1910, A and 1910, B as well as related events and was the source of the editorials. Tacoma events were described in the *Seattle Post-Intelligencer,* meteorite showers and the meteor were reported in the *St. Louis Post-Dispatch.* The sunspot eruption is from *The New York Times* as is the description of the lunar eclipse. These reports are recorded nearly verbatim.

Chapter 5. "Halley's Comet, Celestial Photography, and the Hawaiian Islands Expedition"
Reprint and summary of the expedition's report as it appeared in the American Astronomical Society's Comet Committee Report, 1909-13, in *American Astronomical Society Publications,* 1915.

Chapter 6. "The Comet Observed from Balloons: Up, Up and Away"
The reports of all the ascents were taken nearly verbatim from *The New York Times.*

Chapter 7. "Halley's Comet, Wireless Telegraphy, and the Breakthrough to Modern Radio"
This chapter's material is from *The New York Times* with the exception of the report of Dr. DeForest's demonstration, which is from the *Seattle Post-Intelligencer.*

Chapter 8. "Halley's Comet and the Telescope Trade"
This chapter is taken verbatim from *The New York Times.*

Chapter 9. "Interesting Scientific Experiments"
The report of Professor Birkeland's experiment appeared in *Popular Science.* The description of the Mr. Wilson experiment is from *The New York Times.* Other sources noted in text.

Chapter 10. Astronomer Mary Proctor: "The Lonely Watcher in the Tower"
The reports of Miss Proctor's activities appeared in *The New York Times* from April 26-June 1, 1910. She also wrote about the experiences 16 years later in a book, *Romance of the Comets,* from which some of the information for this chapter comes.

Chapter 11. French Astronomer Camille Flammarion Said, "We May Die Laughing When the Comet Comes"
"Halley's Comet and the Fate of Humanity" appeared in *Busy Man's Magazine* for February, 1910. All other sources are identified in the text.

Chapter 12. "Halley's Comet in the Magazine, *Flaming Sword*"
All material reproduced and discussed is from the magazine *Flaming Sword* for June, 1910.

Chapter 13. "Thou Roosevelt of the Heavens"
Safari information is from Roosevelt's book, *African Game Trails: An Account of the African Wanderings of an American Hunter-Naturalist.*

Chapter 14. "Mark Twain (November 30, 1835-April 21, 1910) and Halley's Comet at Perihelion (November 16, 1835-April 20, 1910)"
All sources identified in the text.

Chapter 15. "'The King Is Dead: Long Live the King:' Halley's Comet, King Edward VII and King George V"
The description of the King's death and related events is found in *The Illustrious Life and Reign of King Edward VII* by W. J. Jackman. The story of events in Hamilton, Bermuda was reported in *The New York Times*. Other sources are cited in the text.

Chapter 16. "What Did the Pope Say about Halley's Comet?"
This account is taken nearly verbatim from *The New York Times*.

Chapter 17. "The Comet in Cartoons and Humor"
William F. Kirk's "Little Bobbie's Pa" and "Our Swede Servant" appeared regularly in the *New York American*. Wex Jones wrote humor for the same newspaper, as did Frank L. Stanton.

Chapter 18. "'The Comet Plain and the Comet Fancy:' Halley's Comet Post Cards"
All sources cited in the text.

Chapter 19. "Halley's Comet and the Ad Men"
All sources cited in the text.

Chapter 20. "Comet Songs and Comet Music: or, 'I Think the Thoughts of Halley, Were Only a Sweet, Sweet Dream'"
Description of the Ziegfeld Follies staging is taken from Charles Higham, *Ziegfeld*. Other sources cited in the text.

Chapter 21. "A Comet Sermon for Children"
Source is cited in the text.

Chapter 22. "Prophesy, Portents, and Edwin Emerson's Book of Terror"
All material quoted and discussed is from Edwin Emerson's *Comet Lore: Halley's Comet in History and Astronomy*.

COPYRIGHTS AND ACKNOWLEDGEMENTS

Contemporary Books, Inc. for permission to quote from *Ziegfeld* by Charles Higham, copyright 1972.

Houghton Mifflin Company for permission to quote from Dixon Wecter's *Sam Clemens of Hannibal.* Copyright 1952 by Elizabeth Farrar Wecter Pike. Copyright renewed 1980 by Elizabeth Farrar Wecter Pike. Reprinted by permission of Houghton Mifflin Company. Also quoted is a passage from unpublished manuscript in MTP, DV 326, "The Mysterious Stranger in Hannibal." Mark Twain Papers, University of California.

Punch for two cartoons, "The Great Amateur" and "'Scuse me, Guv'ner, can you tell me where I can get a view of this 'ere comet?"

The Los Angeles Times for "Camille Flammarion Says We May Die Laughing When the Comet Comes."

PHOTO CREDITS

Special thanks to *Astronomy Magazine,* Astro-media, Inc. for providing most of the photographs included.

Thanks to *American Astronomical Society* for the use of the photograph of Diamond Head Observatory.

Photographic postcard of man astride comet used with permission from *The Gotham Book Mart Collection,* New York City.

Photographic postcard of lady and the comet used with permission from Joe Laufer, *Halley's Comet Watch '86,* Vincentown, N.J.

Photograph of Halley's Comet used with permission from *Lowell Observatory.*

French comic postcards used with permission from the collection of *Sheldon Dobres.*

German comic postcards used with permission from the collection of *Jules Grieten,* Bara Photographic, Inc.